Bjørn Erik Munkvold

Implementing Collaboration Technologies in Industry

Case Examples and Lessons Learned

with contributions from Sigmund Akselsen, Robert P. Bostrom, Bente Evjemo, Jan Grav, Jonathan Grudin, Chris Kadlec, Gloria Mark, Leysia Palen, Steven E. Poltrock, Dominic Thomas and Bjørn Tvedte

With 26 Figures

Bjørn Erik Munkvold, PhD
Agder University College, Department of Information Systems,
Kristiansand 4604, Norway

Series Editors
Dan Diaper, PhD, MBCS
Professor of Systems Science & Engineering, School of Design,
Engineering & Computing, Bournemouth University, Talbot Campus, Fern Barrow,
Poole, Dorset BH12 5BB, UK

Colston Sanger
Shottersley Research Limited, Little Shottersley, Farnham Lane
Haslemere, Surrey GU27 1HA, UK

British Library Cataloguing in Publication Data
A catalogue record for this book is available from the British Library

Library of Congress Cataloging-in-Publication Data
Munkvold, Bjørn Erik, 1962–
Implementing collaboration technologies in industry: case examples and lessons learned/
Bjørn Erik Munkvold.
p. cm. – (Computer supported cooperative work, ISSN 1431-1496)
Includes bibliographical references and index.
ISBN 1-85233-418-5 (alk. paper)
1. Information technology–Management. 2. Communication in organizations. 3. Teams in the workplace. 4. Business networks. 5. Computer networks. I. Title. II. Series.

HD30.2.M88 2002
658.4'038–dc21 2002026867

CSCW ISSN 1431-1496
ISBN 1-85233-418-5 Springer-Verlag London Berlin Heidelberg
a member of BertelsmannSpringer Science+Business Media GmbH
http://www.springer.co.uk

Typesetting: Gray Publishing, Tunbridge Wells, UK
Printed and bound at the Athenæum Press Ltd., Gateshead, Tyne & Wear
34/3830-543210 Printed on acid-free paper SPIN 10790615

Contents

Preface

The focus on information and communications technologies (ICT) as a medium for collaboration has gradually increased over the last two decades. Since the early days of groupware in the 1980s, when this technology still had a somewhat exotic aura, collaborative functionality has become embedded in many standard tools available in the workplace. The advent of the Internet and the Web has represented a radical development in infrastructure for distribution and access to collaborative services. Collaboration support is now often taken for granted; it has become an integral element of the many current strategic initiatives such as knowledge management, e-Commerce, process improvement, virtual teams and global collaboration.

However, while the practitioner press has touted the dramatic potential of this new technology for supporting organisational transformation, the academic research literature has reported numerous failures in these efforts. This has clearly highlighted how the process of actually realising the anticipated benefits from collaboration technology can often be arduous and time-consuming. Some of these problems can be ascribed to the relatively immature stage of these technologies, and as the technologies and related practices are developing the number of success stories reported is also increasing. Nevertheless, the goal of applying these technologies to increase and improve collaboration remains a challenge for many companies.

The focus of this book is on how companies are implementing and using different types of collaboration technology. By reporting industry experience and discussing this in relation to previous field studies on collaboration technology implementation, the goal of this book is to provide practical insight of value to both practitioners and academics. The target audience for this book thus includes managers at all levels, project managers, process owners, IT personnel, IT implementation team members, consultants, vendors, as well as researchers and educators.

I first became interested in collaboration technologies in 1993, while working for a Norwegian telecommunications company. Back then, although some organisations had started to explore Lotus Notes, most companies were yet to become familiar with collaboration technologies (or groupware, the common term then). Although working as a systems developer, e-mail was not yet included in my repertoire of work tools, and Internet/web was not yet easily available to the "general public".

It was in this context that I started out on my PhD research, studying how small and medium-sized enterprises could use "modern" ICT to form virtual companies that could compete with larger enterprises. From originally focusing on the evaluation aspects of this technology implementation, and trying to measure the benefits from this, it soon became clear that there were many barriers in this implementation process that represented threats to realising these benefits. When expanding my field studies to other organisations, I reached similar conclusions: organisations were experiencing problems in the implementation of these technologies, often resulting in project delays, ineffective use and frustrated users. In my doctoral thesis (Munkvold, 1998a), I presented an analysis of issues that were found to influence the different stages in the implementation projects.

Since then, my main research interest has been in the process of implementing different collaboration technologies in industry. In trying to follow the development in this area, I have made the following observations. Despite a growing body of research literature, there are few sources that provide practical advice related to the implementation of collaboration technology. Many of the field studies in this area do an excellent job of illustrating the complex processes involved in the implementation of this technology. However, being mainly targeted towards an academic audience, it is often hard to extract the practical implications from these studies. For practitioners looking for advice on how to conduct such implementation projects, it is not enough to be told that this is a complex and sometimes "unmanageable" process – some guidelines are needed as well.

The more practitioner-oriented sources that do purport to provide solutions suffer from being fairly superficial, treating "groupware implementation" as a unified process without differentiating between different forms of technology and different organisational settings. Finally, in much of the academic and practitioner literature there is a lack of explicit focus on how the characteristics of a particular collaboration technology may influence its implementation within an organisation.

Taken together, these observations formed the key point of departure for this project. Rooted in my former industry background, I wanted to write a book that could offer practical advice on the implementation of different collaboration technologies based on industry experience. Further, I wanted to present this experience in a format so that it can be of interest and use to both practitioners in the field and researchers working in this area. Thus, this book is a response to the general call for more focus on practical relevance in academic research, both regarding content and presentation style and format (Benbasat and Zmud, 1999). However, trying to keep the material both practical and yet relevant for an academic audience represents a challenging balancing act.

I soon realised that in addition to my own field-based research, it would strengthen this project to be able to also include lessons learned from colleagues working in this area. I therefore approached a selection of authors whose work and experience I knew was highly relevant to this project and invited them to contribute with a chapter. Many of these authors have a combined industry and academic background, and thus were ideally suited for meeting my dual target audience of practitioners and academics.

Another challenge in writing a book on collaboration technologies is the rapid technological development, that makes this whole field a moving target as new technologies and related applications and concepts emerge. Yet, the main emphasis in this book is not on the technology itself, but on the process of introducing this in organisations and making it work. As will be illustrated by many examples throughout the book, while the technologies continue to develop in functionality and applications, many of the experiences related to the implementation of different types of collaboration technology remain the same.

Several people have contributed in the making of this book. First, I would like to thank the contributing authors for accepting my invitation to share their experience from the field. This really constitutes the essence of the book. The editorial process related to the development of these chapters also stimulated some fruitful discussion that had important bearings on the rest of the book. I will especially like to thank Bjørn Tvedte and Jonathan Grudin for their reflective comments.

A great part of this work was conducted while I was a Visiting Fellow at the University of New South Wales in Sydney, Australia. I am grateful to the faculty of the School of Information Systems, Technology and Management for providing me with good working conditions, and making my stay a fruitful and pleasant one. Good on you, mates! A special thanks here goes to Geoffrey Dick for taking care of all practicalities. I am also grateful to my employer, Agder University College, for funding this research visit, and especially to Department Chair Carl Erik Moe for his support in arranging this.

My research associates and mentors, Professors Rob Anson, Deepak Khazanchi and Maung K. Sein provided useful comments in the finalisation of the manuscript, also cleaning up my worst cases of "Norwenglish". The Springer representatives, Rosie Kemp, Karen Borthwick and Melanie Jackson, exerted the gentle but yet persistent pressure needed to keep me going. Nice working with you! Finally, a special acknowledgement to Professor Bob Bostrom, whose guidance and support throughout this project was key in getting it done.

I dedicate this book to my wife and children, whose affection and support is my constant inspiration and lifeline.

Bjørn Erik Munkvold
Kristiansand, Norway
May 2002

List of Contributors

Sigmund Akselsen
Senior Research Scientist
Telenor Research and Development
E-mail: Sigmund.Akselsen@telenor.com

Robert P. Bostrom
L. Edmund Rast Professor of Business
Management Information Systems
Department
Terry College of Business
University of Georgia
E-mail: BBostrom@terry.uga.edu

Bente Evjemo
Research Scientist
Telenor Research and Development
E-mail: Bente.Evjemo@telenor.com

Jan Grav
Adviser
Telenor Research and Development
E-mail: Jan.Grav@telenor.com

Jonathan Grudin
Senior Researcher
Adaptive Systems and Interaction Group
Microsoft Research
E-mail: jgrudin@microsoft.com

Chris Kadlec
MIS PhD Student
Management Information Systems
Department
Terry College of Business
University of Georgia
E-mail: ckadlec@terry.uga.edu

Gloria Mark
Assistant Professor
Department of Information and
Computer Science
University of California, Irvine
E-mail: gmark@ics.uci.edu

Bjørn Erik Munkvold
Associate Professor
Department of Information Systems
Agder University College, Kristiansand,
Norway
E-mail: Bjorn.E.Munkvold@hia.no

Leysia Palen
Assistant Research Professor
Department of Computer Science
University of Colorado at Boulder
E-mail: palen@cs.colorado.edu

Steven E. Poltrock
Technical Fellow
The Boeing Company
E-mail: steven.poltrock@boeing.com

Dominic Thomas
MIS PhD Student
Management Information Systems
Department
Terry College of Business
University of Georgia
E-mail: dominict@terry.uga.edu

Bjørn Tvedte
Staff Engineer
Corporate Services
Statoil
E-mail: btve@statoil.com

Part I

Overview and Current Trends

1 Introduction

Bjørn Erik Munkvold

1.1 Background and Scope

The purpose of this book is to present current experiences from industry related to the organisational implementation and use of collaboration technologies. Collaboration technology is defined here as all types of information and communication technologies that enable collaboration at various levels, from two persons co-authoring a document to inter-organisational collaboration where several companies are engaged in common tasks. Examples of such technologies include video and desktop conferencing, knowledge repositories, workflow management systems, online meeting schedulers and electronic meeting support systems. The number of organisations currently adopting these technologies is increasing rapidly. Collaboration technology constitutes the enabling infrastructure for key elements in business strategy today, such as knowledge management, process improvement, virtual teamwork, global collaboration, and e-Learning. Business experts such as the Gartner Group argue that "collaborative commerce" (c-Commerce) will represent the next stage in the development of e-Business applications, expanding beyond mere transaction processing to also include dynamic collaboration among employees, business partners and customers (Bond et al., 1999).

The term "organisational implementation" incorporates all activities related to deployment and adoption of a new technology, namely requirements specification, acquisition and/or design and development, installation, training and internalisation of routines for effective utilisation. Despite increasing diffusion of different types of collaboration technology in organisations, little formalised knowledge exists on how to implement these technologies in organisations to realise the potential benefits. Empirical studies to date have illustrated the complexity involved in implementing collaboration technology, involving a dynamic interplay between technological, organisational and project management aspects. However, much of this research does not address the practical implications from these findings. There is a gap in the literature for sources offering solutions and guidelines for practice based on solid empirical research, also taking into account the contextual contingencies characterising projects in the "real world".

This book is a contribution to fill this gap, by giving a detailed presentation of the experiences from implementing different types of collaboration technology in selected organisations. The focus here is on the strategies applied in each case for overcoming the various obstacles in the implementation process, and the measures taken to enable effective use. Rather than presenting "quick fix" solutions, the studies give careful consideration to the various barriers in the process, and the possible solutions for overcoming them. The case studies presented reflect the variation in collaboration technologies, applications and organisational setting characterising practice today. By including contributions from industry practitioners, first-hand experience from this type of implementation project is presented, together with analyses from outside observers.

The book also includes a review of previous empirical research on implementation of collaboration technology. While previous reviews have focused mostly on one or a few types of collaboration technology, the review presented here covers implementation research related to the entire spectrum of collaboration technologies. Based on this review, a taxonomy of implementation factors is developed for categorising the empirical findings. This taxonomy is used to structure the analysis of the case studies presented in the book, and can also serve as the basis for further research on collaboration technology implementation. The book thus constitutes a valuable reference for both practitioners and academics. In addition to offering useful experience and insight for practitioners, the book can be seen as a response to the call for more practical research, aimed at increasing the sensitivity of academics to "what is going on in practice" by describing, evaluating and interpreting practice (Markus, 1997).

1.2 Structure of the Book

The book consists of three parts. Part 1 (Chapters 1–4) gives an introduction to the area of collaboration technologies and previous research on implementation of these technologies. Chapter 2 provides an overview of different collaboration technologies, introducing a framework for classifying these. Chapter 3 presents a comprehensive review of previous research on implementation of collaboration technologies, with main emphasis on the practical findings from field-based research. Based on these findings, Chapter 4 introduces a taxonomy of implementation factors for collaboration technologies.

Part 2 (Chapters 5–10) contains six case studies of organisational implementation and use of different collaboration technologies. Together, these cover a wide range of technologies, organisational contexts and practical experience. Several of these chapters are written by invited authors, who are leading researchers in the field and have a close affiliation to industry.

Chapter 5, co-authored with Bjørn Tvedte (Statoil IT), describes the organisational process related to implementation of a portfolio of collaboration technologies in Statoil. This case gives a unique insight into how implementation and use of different collaboration technologies is interrelated, calling for an integrated perspective on this implementation. The chapter also presents the on-going strategic process in Statoil for defining the next generation of collaborative infrastructure in the company.

Chapter 6 presents the experiences from implementation and use of collaboration technologies in Kværner, a multinational engineering group. The first part of the chapter discusses different barriers related to implementing a global network infrastructure in Kværner. The second part presents experiences from Kværner's current use of collaboration technologies, and further challenges in their process of becoming a global, virtual engineering company.

In Chapter 7, by Steven Poltrock (Boeing) and Gloria Mark (University of California, Irvine), an in-depth account is presented of the implementation of data conferencing services in the Boeing company. Over more than a decade, the company has gradually developed a robust data conferencing service that today is widely used within the company for supporting distributed meetings. The chapter presents various technical and organisational challenges in this process, and the strategies applied for meeting these challenges.

Chapter 8, by Leysia Palen (University of Colorado, Boulder) and Jonathan Grudin (Microsoft Research), analyses the adoption and current use of online calendar tools in two major companies in the IT industry, Sun and Microsoft. By contrasting these findings with similar studies conducted in the mid-1980s, the chapter identifies important factors leading to the universal adoption and use of this technology within these companies today.

Chapter 9, by Bente Evjemo, Sigmund Akselsen and Jan Grav (Telenor R&D), presents experiences accumulated by researchers in Telenor, a Norwegian telecommunications company, through their extensive field trials of developing collaborative services for different sectors. Unlike the other case studies, these trials have mainly been conducted within small and medium-sized enterprises, and in the public sector. The experiences are summarised as "ten commandments" for implementing sector-specific, collaborative tools.

The case study in Chapter 10, by Robert P. Bostrom, Chris Kadlec and Dominic Thomas (University of Georgia), represents yet another application area and organisational context: a joint e-Learning project between the University of Georgia, Terry College of Business, and PricewaterhouseCoopers North American Consulting Group. The chapter describes the development and implementation of a MBA programme tailored for the PwC consultants. Through using an e-Learning infrastructure including various tools for virtual team support, the students are able to complete the programme in normal time while still working for their employer. The chapter presents important factors and guidelines leading to successful implementation of this type of e-Learning programme.

Part 3 (Chapters 11 and 12) integrates the findings from the case studies in Part 2. Chapter 11 presents a cross-case comparison and discussion, structured according to the taxonomy of implementation factors presented in Chapter 4. Based on this analysis, Chapter 12 finally presents practical implications and guidelines from this research, and some implications for further research on implementation of collaboration technologies.

2

Collaboration Technology: Overview and Current Trends

Bjørn Erik Munkvold

2.1 Introduction

Chapter 1 presented a definition of collaboration technology. This chapter gives an overview of the main categories of collaboration technologies, and provides examples of applications of these technologies. Rather than providing an exhaustive guide to the area, the aim here is mainly to give the reader sufficient background for understanding the technological context of the case studies presented in Part 2 of the book. Thus, the description of the technologies focuses more on functionality and possible applications rather than technical issues.

The next section provides a brief historical account of the development of this field. This is followed by a typology for classifying the various technologies, and a brief description of each category. The chapter ends with a summary of some major trends characterising the further development of this area.

2.2 Brief Historical Background

The term collaboration technology is closely related to the terms *CSCW* (Computer Supported Cooperative Work) and *Groupware*. The term CSCW was coined during a workshop at Digital in 1984, where a group of researchers focused on how IT could support collaboration (Grudin and Poltrock, 1997). In general, the limited success from the office automation movement in the 1980s spurred an increasing focus on how IT could evolve from a single-user tool to become a medium for collaboration among groups of people. This required more detailed analysis of the actual collaborative work practices enacted by organisational employees. The increasing merger between computer and telecommunications technologies also opened new possibilities for communications support. One of the earliest works applying the term groupware is the book *Groupware: Computer Support for Business Teams* (Johansen, 1988), which presented suggestions on how to apply this new technology for supporting team collaboration.

The origins of this field, however, can be traced back to the 1960s and the work of Douglas Englebart and colleagues at the Stanford Research Institute. Although not widely acknowledged, these researchers actually laid the groundwork for many of the current features of modern computer technology and applications, such as word-processing and hypertext. They also ran a data conferencing facility, operated by a mouse and pen-based interface.

Since the late 1980s, several CSCW and groupware forums have been established where researchers and industry practitioners meet for conference presentations, tutorials and general networking. These forums are truly multi-disciplinary, incorporating the whole spectrum from "hardcore" computer scientists to "softcore" sociologists and psychologists, but all sharing an interest in how IT may support collaboration. An overview of major forums is provided in Appendix A.

Reflecting the relatively early stage of this area, the main focus in these conferences has been the design and modelling aspects of collaborative applications. However, as more technologies and applications have diffused into industry, there is also an increasing number of field studies on the implementation and use of these technologies in organisational settings.

In general, the application of these technologies has broadened from originally supporting the workgroup or team level, to organisation-wide applications such as document management systems or knowledge repositories. The term "collaboration technology" is better able to represent this wider focus, and the "groupware" term has gradually faded. With the rapid change in this area the term collaboration technology will almost certainly be of a transient nature as well. Already, it is common to see the same technologies being referred to as *Knowledge Management Technologies*, or technologies supporting *Digital Collaboration*, *e-Collaboration* and *c-Commerce*.

A more comprehensive review of the history and contents of the CSCW and groupware research area can be found in Grudin (1994a) and Grudin and Poltrock (1997). The following section will discuss different ways of classifying collaboration technologies and applications.

2.3 Classifying Technologies and Applications

The broad range of collaboration technologies and applications has resulted in several taxonomies and classification frameworks. The most

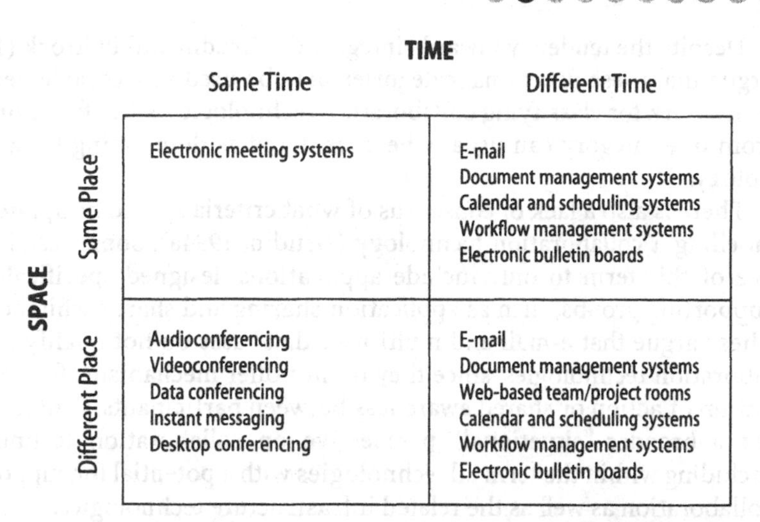

Figure 2.1 Time-space matrix for classifying collaboration technology

influential of these has been the time/space taxonomy first presented by DeSanctis and Gallupe (1987). This organises the collaboration technologies into a four-cell matrix, according to the time and geographical space of the collaborative interaction, as illustrated in Figure 2.1.

Same time interaction is also commonly referred to as *synchronous* interaction, while different time is termed *asynchronous*. Figure 2.1 includes typical examples of technologies in each cell, and these will be described in the following sections. As shown in the figure, however, different technologies may support interaction in more than one mode.

Although useful as a simple tool for classification of technologies, this typology has its limitations. Organisational work is seldom restricted to one of these cells, but rather involves combinations of different time/place interactions (Grudin and Poltrock, 1997). As a consequence, an important trend for collaboration technologies is the integration of different functionality and services within the same product or through providing seamless integration between products. This integration can thus support multiple modes of collaboration, or so-called "anytime, anyplace" collaboration. This requires support for the following types of collaborative tasks:

- *communication*; interpersonal communication through audio, text, video, etc.
- *information sharing*; creation and manipulation of shared information objects
- *coordination*; managing interdependencies between participants and their activities.

Despite the tendency towards integration, Grudin and Poltrock (1997) argue that these functional categories may be used as a complementary framework for classifying collaboration technologies, since functionality from one category can usually be considered as dominating for a technology.

There is also a lack of consensus of what criteria should be applied for labelling a collaboration technology (Grudin, 1994a). Some restrict the use of this term to only include applications designed specifically for supporting groups, such as application sharing and shared whiteboards. These argue that e-mail and multi-user databases do not qualify as collaboration technologies, since they do not offer mechanisms for coordination of action or shared awareness between participants. Others argue for a broader "situational" perspective on collaboration technology, including within the term all technologies with a potential for supporting collaboration as well as the related infrastructure technologies.

This book takes on the latter perspective on collaboration technology. Thus, rather than implying specific criteria to restrict this area, the focus will be on various technologies that are being used for collaborative purposes. Table 2.1 presents a simple classification framework, building on the functional model presented above, and comprising five categories of technologies.

In addition to the three categories from the functional model, this framework includes two additional categories: meeting support technologies and integrated products. Meeting support technologies could also be treated as part of the shared information space category (Grudin and Poltrock, 1997), but is here included as a separate category due to the special application area of organisational meeting processes targeted by this technology. Integrated products combine features from several of the other categories.

Table 2.1 Categories of collaboration technologies

Main categories	Examples of technologies
Communication technologies	E-mail Instant messaging Audio/videoconferencing
Shared information space technologies	Document management systems Web-based team/project rooms Data conferencing/application sharing Electronic bulletin boards
Meeting support technologies	Electronic meeting systems
Coordination technologies	Workflow management systems Calendar and scheduling systems
Integrated products	Collaboration product suites, integrated team support technologies, e-Learning technologies

An important prerequisite for the use of the different technologies in Table 2.1 is a functional network infrastructure. The following list includes examples of such technologies:

- LAN (local area network)
- WAN (wide area network)
- ISDN (integrated services digital network)
- DSL (digital subscriber line)
- ATM (asynchronous transfer mode)
- Internet
- Intranet
- Extranet

The quality of these services regarding performance, stability and compatibility can have vital impact on the implementation and use of a collaboration technology. Although not providing any collaborative functionality by themselves, the issues involved in implementing network infrastructure technologies are often similar to those in the implementation of collaboration technologies and services. This is especially prevalent in cases where such network technologies are implemented as a first step towards a collaborative infrastructure (Evaristo and Munkvold, 2002) enabling virtual collaboration among different companies or organisational units (Munkvold, 1999). Several examples of this are provided in the case studies in Part 2 of this book. In general, performance and capacity of these network technologies are constantly being improved through new technologies such as broadband Internet. However, new application areas of these technologies involving increased linking of different organisations continue to pose new challenges regarding interoperability and security.

In the following sections, a brief introduction to each category in Table 2.1 is presented, including main functionality, current status of diffusion and application, challenges and future trends. In general, the focus will not be on specific products although some examples are mentioned. Appendix B provides further product examples. The inclusion and specific mentioning of product names in this book does not imply any superiority or specific recommendations of these products compared to others. Trademark is implied when referring to product names.

2.3.1 Communication Technologies

This category includes both asynchronous and synchronous technologies that support interpersonal communication across geographical distance. These technologies mainly function as electronic communication channels, and do not include functionality for processing information.

The major technologies in this category are e-mail, instant messaging, audio- and videoconferencing.

E-mail

E-mail is by far the most widespread electronic communication technology today, both for business and personal use. The simple and intuitive use of this service combined with the benefits of efficient one-to-many communication has contributed to this success. Some also argue that e-mail communication leads to better decision processes, in that people are able to provide more reflected input than in synchronous, face-to-face communication. In addition, e-mail also plays a vital role providing messaging services for coordination tools such as workflow, and calendaring and scheduling.

Interoperability of different e-mail systems has been and still is an issue, and many large organisations such as Boeing, Statoil and Kværner represented in the case studies in this book have struggled with integrating address directories etc. for different e-mail systems used throughout the organisation. Although technical standards such as X.400 and SMTP/MIME now exist for seamless transfer of messages between different e-mail systems, efforts of standardising on one single system in an organisation may also face resistance from different user groups.

Still, the main challenges to effective use of this technology today are related to e-mail etiquette (such as avoiding "flaming"), and establishing routines for avoiding redundant communications (such as individual e-mail copied to all employees) and information overload. Virtual teams relying solely on electronic communication often develop procedures for effective communication such as using the Subject field for classification of content and version control, and "protocols" for acknowledging messages to create closed "communication loops". More advanced e-mail services include the use of intelligent agents performing operations based on incoming e-mail and filtering mechanisms for sorting mail and deleting unwanted messages.

Instant Messaging

Even though e-mail is the most widespread communication tool today, it has some limitations. When sending an e-mail, you do not know if the receiver is online, and the asynchronous mode of this tool makes it difficult to have a running conversation with this person. Instant messaging (IM) combines the real-time advantages of a phone call with the convenience of e-mail. With IM you can see whether a person is online, and exchange text messages in near real-time. Most IM-products also include functions for establishing chat rooms with friends and co-workers, exchanging files, and conducting audio- or videoconferences. IM therefore also serves as a medium for coordinating interaction, launching other communications services after checking the communication partner's presence and availability.

This technology is already widely popular for social interaction among home users. The number of business users of this service is growing rapidly. Public IM clients are today provided free by large companies such as Microsoft, America Online and Yahoo. In the wait for a universal IM protocol, however, users of these different services are not able to communicate with each other. There are also security limitations in the use of these public IM channels, in that user IDs and passwords are submitted in plain text and are unencrypted.

Audioconferencing

Audioconferencing is another intuitive service frequently used for organisational meetings. Initiating these conferences is also becoming increasingly simple, as telecommunications companies provide instant connection through the web. An increasing number of vendors also offer web-based conference services based on Internet telephony, often integrated with data conferencing services. However, the sound quality here may still represent a limitation.

With communication restricted to one channel only, effective use of audioconferencing requires some form of moderating to ensure a smooth flow of communication, and participation and engagement from everyone in the meeting. Reflexive listening (rephrasing participants' statements as acknowledgment) and questions targeted to specific participants are some possible strategies here.

Videoconferencing

Videoconferencing is another example of an "old" collaboration technology that has been available on the market since the 1970s. Most organisational installations of videoconferencing typically include a dedicated video room (studio) with a large screen projecting the speaker and participants at remote sites, but there also exist mobile (roll-about) units that can be transported between different meeting rooms. The potential cost savings by using this technology are easily identifiable, in the form of reduced time and money spent on travels. Still, although many organisations today apply this technology on a regular basis, it has not achieved the same widespread use as e-mail. The costs related to "traditional" videoconferencing equipment have been considered as relatively high, and functionality has also been restricted as a result of limited integration with other technologies for supporting document sharing, etc.

This situation is changing with the advent of PC-based desktop videoconferencing. Combining audio and videoconferencing, these systems enable employees to participate in distributed meetings from their offices with images of the other participants displayed as "talking heads" on the screen. By also combining these functions with information sharing capabilities through application sharing, you get a complete conferencing environment available from your desktop. These systems are termed *desktop conferencing systems* and will be described further under the category of Integrated products.

Desktop videoconferencing is also used for organisational broadcast of presentations such as management announcements, etc. Compared to traditional TV technology this offers the possibility for two-way communication, enabling questions and comments from the audience, quick user polls and so on, as part of the broadcast.

The benefit from desktop videoconferencing has been restricted by the limited resolution of the picture combined with difficulties of making eye contact (Grudin and Poltrock, 1997). However, the quality and performance is constantly improving while the price is going down. Videoconferencing systems today can be run on phone lines, ISDN and Internet/web, and the rapid diffusion of broadband technologies will increase capacity further.

Still, many seem to prefer meeting face-to-face, even though this implies spending time and resources on travelling. The value added by video for supporting communication is also an issue for debate. Perey (1997) puts emphasis on the potential for conveying sincerity and authority and contributing to the bonding of people, while others question the value added by including this visual information (Line, 1998; Mark et al., 1999). In general, the potential importance of video seems to be greater during the early stages of teambuilding (as a substitute for face-to-face meetings when these are not possible) than in later collaboration among parties that know each other well.

2.3.2 Shared Information Space Technologies

Shared information space technologies is the common term used for technologies supporting collaborative work related to the production and manipulation of information objects such as documents and drawings, and for creating virtual interaction spaces such as electronic bulletin boards and discussion lists. This category includes both synchronous and asynchronous technologies. Four types of technologies will be addressed here: document management systems, web-based team/project rooms, data conferencing and electronic bulletin boards.

Document Management Systems

Document management systems is a class of products that support the creation and electronic archiving of documents. Based on a dedicated document management server, this involves the administration of access control (read, write access, etc.), concurrency control (check in and check out of documents to avoid conflict between users) and version control (ensuring that all users work on the latest version). Few document management systems maintain information on the structure of the document objects in the database. Instead, document searches are based on document meta-data such as author, date, version number, user-supplied keywords and access permissions. Getting the users to enter this meta-data may represent a challenge to effective utilisation of these systems (Grudin and Poltrock, 1997).

In addition to a growing number of specialised document management systems on the market, this functionality is also offered as part of integrated products such as Lotus Notes and MS Exchange. In general, document management functionality is increasingly integrated with other categories of systems, such as workflow applications and enterprise resource planning systems (ERP).

There is a growing focus on document management systems in organisations today for managing the vast amounts of information produced. The term Content Management (CM) represents the latest development in this area, focusing on strategies for applying IT to support the entire information management in an organisation. Ideally, these systems will provide access to all relevant information content in the organisation, regardless of format or media, through effective search capabilities (Votsch, 2001). Although vendors currently market CM solutions, a complete CM infrastructure is dependent on integration of several technologies. Information portals are an important element in this infrastructure, offering personalised access to relevant information for company employees ("employee portals") through a web-based interface. The interest in these technologies is also spurred by the current focus on knowledge management (KM), as document management systems provide the functionality to serve as knowledge repositories in a KM architecture.

Web-based Team/Project Rooms

There is an increasing number of web-based services available that offer free creation and use of document repositories within a limited space. While the functionality of these does not cover features such as version and concurrency control, and lacks the scalability of "complete" document management systems, they may still provide useful support for temporary workgroups and ad hoc projects. They are also well suited for supporting inter-organisational collaboration, as they do not require any common infrastructure beyond the Internet/web. Many of these offer other tools integrated with the document repositories, such as discussion lists, calendars and project management tools. These products are described further in the Integrated products category.

Data Conferencing

Data conferencing supports synchronous communication and information sharing through exchange of text-messages (chat), shared whiteboard and application sharing. A shared whiteboard is a multi-user graphics editor in which users can work (draw, type, annotate, etc.) simultaneously. A special type of shared whiteboard is available for use in conference rooms. It consists of a large, touch-sensitive electronic whiteboard that can be used for sharing and editing computer-generated and hand-written information through cordless infrared pens (Blundell, 1997). Examples of this type of system are SMART Board and LiveBoard.

Application sharing enables two or more people to work together in any program running on only one of the participants' machines. The users each have their own cursor and may take turns on controlling the application. In addition to supporting distributed meetings, this functionality is becoming increasingly used for training and demonstration purposes where the instructor may share his or her screen with the participants. Similarly, the technology is used for running presentations at multiple sites, with the presenter distributing his material (Powerpoint slides etc.) to workstations at different locations. Interestingly, the Collaborative Strategies Group predicts data conferencing technology as the next major growth area among collaborative technologies (Collaborative Strategies, 2000). A product example in this category is Microsoft NetMeeting, that is "bundled" with the Windows operating system. Chapter 7 presents the experiences from the implementation and use of this technology for supporting distributed meetings in the Boeing company.

Electronic Bulletin Boards

Electronic bulletin boards are shared virtual information spaces that support threaded discussions by using a tree-structured database. Unlike e-mail this makes it possible to maintain a structure of the asynchronous communication in the form of main topic, responses and responses to responses. Different variants of bulletin boards are discussion lists, news groups and public folders, and these may be run on the Internet, intranets

or company LANs. Lotus Notes is also an example of a technology based on this threaded structure.

2.3.3 Meeting Support Technologies

Meetings are the most common form of interaction in companies; however, a range of factors contribute to make traditional meetings ineffective. Only one person can speak at a time, leading to different problems of *production blocking* (Ellis et al., 1991) such as people losing their string of thought, lack of attention, etc. This form of interaction also often results in a few people dominating a meeting, based on their formal position and/or verbal skills, while subordinates or people of a more modest nature are not fully heard. All this may lead to valuable input and participation being lost. Further, the outcome of the meeting is often not properly documented and the follow-up actions remain unclear, thus requiring new meetings and so on.

The question of how to apply IT for making meetings more effective has been on the research agenda for more than three decades. Originating in academic research settings, a class of technologies termed electronic meeting systems (EMS) has been developed with the aim of reducing the negative effects described above, by offering the following functionality:

- *parallel input* through providing each participant with a workstation (e.g. a PC or a laptop) for input of ideas and further processing of these;
- *equal participation* through full anonymity of all keyboard entries, thus enabling each meeting attendant to participate on an equal basis regardless of formal position or verbal skills;
- *full meeting documentation* through automated generation of meeting reports, including roster, agenda, all input generated from each participant, categorisation and ranking of ideas, etc.

EMS products vary in functionality, but the basic features involve agenda specification, procedural guidance, electronic brainstorming, recording and storage (also referred to as group memory), idea organisation, alternative analysis and reporting. The market leader for this type of system has so far been GroupSystems.com, with around 1700 installations worldwide. A growing number of web-based systems are also available, offering flexible access through web-browsers. Some examples of these are listed in Appendix B.

So far, EMS have mostly been applied in specially equipped meeting rooms, with participant workstations arranged in horseshoe formation facing a large screen or Liveboard for projecting the participants' input and common focus during idea organisation, etc. Some examples of such meeting rooms can be viewed at the GroupSystems website (www.groupsystems.com). In co-located settings, these meetings normally apply a mix of verbal discussion and electronic interaction through the workstations. To avoid obstructing visual contact among the participants, the screens may be submersed into the desks, or laptops may be used as individual workstations.

Another important part of the infrastructure for electronic meetings is the use of a meeting facilitator (Anson et al., 1995; Jessup and Valacich, 1993). While traditional meetings also may use facilitators specially trained in different techniques for supporting creativity and decision processes in groups (such as "sticky dots"), the use of a facilitator is an absolute necessity for running an effective, co-located electronic meeting. The facilitator here takes part in the pre-planning of the meeting with the meeting owner (the person calling the meeting), defining the agenda and mapping the various tools to the different activities in the meeting. During the meeting the facilitator guides the participants through the different meeting activities, controlling the start and end times for use of each meeting support tool, facilitating verbal discussion and ensuring that the group stays focused towards the intended outcome of the meeting. This role is challenging and several organisations prefer to split this task between two individuals, where one is responsible for the technology part and the other for facilitating the meeting process. The recruitment and training of facilitators represents a key challenge to effective utilisation of this technology in an organisation.

There is also a growing use of this technology for supporting distributed meeting processes. Anson and Munkvold (2002) describe how Statoil, a Norwegian oil company, applies this technology for supporting different combinations of time and place modes, running GroupSystems on the company LAN. Distributed meetings like these involve different meeting processes compared to co-located meetings regarding facilitation and the nature of interaction between the participants. The requirements for distributed facilitation services are an area of increasing interest, and several companies are starting to emphasise this role (Mark et al., 1999).

2.3.4 Coordination Technologies

Coordination can be defined as the act of managing interdependencies between activities performed to achieve a goal (Malone and Crowston, 1990). Although all collaboration technologies need some sort of coordination mechanism to support task and work interdependencies, two main technologies are usually included in this category. These are workflow management systems and online calendars and meeting schedulers.

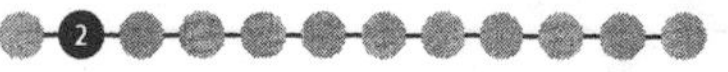

Workflow Management Systems

Workflow management systems (WfMS) is the common term used for technologies that support the automation of work processes by routing information among the different actors according to a predefined (or ad hoc) sequence representing the process. The goal is to provide the right person with the right information at the right time. Typical applications are loans processing or insurance claims handling, where the case (loan, insurance claim) is distributed to different employees responsible for carrying out the various tasks in the handling sequence. As each case handler completes his/her task (such as loan approval) the necessary information is passed on to the person responsible for the next stage in the process. The system provides each participant with a job queue, and notifies them of new entries requiring their attention. It is also possible to assign priorities to the different tasks.

When defining the work processes in the system, the different actors are specified by roles instead of specific employees. This avoids dependence on specific persons, and reduces problems in case of sick leaves or other changes in personnel. This technology also offers the possibility for detailed control and supervision of the work process, by monitoring and providing status on the execution of each task, time used at each stage and so on. The WfMS are often integrated with other company information systems, such as OCR systems for scanning paper documents, document management systems, customer databases and accounting systems, to provide access to the necessary information at each stage in the work process.

In general, the potential benefits from WfMS can be summarised as:

- increased productivity through faster distribution of information;
- improved quality through consistent and standardised processes;
- improved customer service through faster response time;
- increased process control through monitoring and tracking ability;
- more integrated and effective use of IT resources.

Although most of the applications of this technology are found in administrative domains, there are also many examples of engineering industries applying this technology. For example, WfMS play a crucial role in the Boeing company for managing and keeping track of the thousands of work operations and parts used in building and maintaining their aircrafts. In general, the application of WfMS may span from ad hoc processes, such as the routing of a special customer request, through to automation of core production processes such as loan approvals (Marshak, 1997).

The architecture and functionality of WfMS may vary. Marshak (1997) describes three alternate client–server architectural models: the mail-based model, the shared database model and the client–server database model. See also Stohr and Zhao (2001) for a good overview of WfMS frameworks and architectures. While some systems offer support for the entire process from business process design to production of this process, others only contain the "workflow engine" maintaining the workflow. In general, the lack of standardisation of interfaces between the different modules in a WfMS has been a problem, but the Workflow Management Coalition is working on arriving at a set of common standards (WfMC, 2002).

WfMS are closely related to business process reengineering (BPR) and often constitute an important part of the technology infrastructure needed to implement the new processes. Many workflow products also include graphical tools for designing the new work processes. A business process is typically defined as a sequence of activities or tasks, that may be conducted in parallel and/or sequence. A process can also be split into different sub-processes based on predefined conditions, and may span several organisational units (departments, etc.). However, workflow tools need not necessarily be implemented as part of a BPR project, but can also be used for automating existing work processes to make these more efficient.

Despite the potential benefits summarised above, the market for WfMS has been slow and workflow applications are still not very wide-spread in use. Marshak (1997) points to several factors acting as barriers to the organisational uptake of this technology. First, many organisations tend to focus on large scale re-engineering when implementing this technology. The time and money required to undertake this type of project may scare companies from embarking on this. Further, although several workflow products today offer a great deal of flexibility in design and adaptation of the work processes, the handling of exceptions to the standardised work processes still represents a challenge to the utilisation of these systems. Modifying workflows may be especially difficult when the WfMS is closely integrated with other IT systems in the organisation. In today's fast changing environments with frequent organisational changes, the "fixed representation" of work processes implemented in WfMS may thus create concerns in companies. Further, although management may perceive the possibility for increased process control as a major benefit, from the users' perspective the increased monitoring and accountability may create a sense of discomfort and "Big Brother".

The following trends can be identified for workflow technology:

- increasing focus on the Internet/web as the platform for WfMS – this means that participants in the work processes can access their job queues and perform the tasks through a web browser;

- increased focus on inter-organisational workflows as part of the trend on business-to-business (B2B) e-Commerce and integration of value chains;
- integration of workflow functionality in other organisational systems such as document management and ERP systems.

Together, these trends may contribute to spread the use of this technology further.

Online Calendars and Meeting Schedulers

The second major category of coordination technologies is online calendaring and scheduling systems. Although often referred to as one technology, this actually comprises two different functionalities. Calendaring involves the placement and manipulation of data on a calendar, while scheduling involves the communication and negotiation between calendars for such data placement (Knudsen and Wellington, 1997). In other words, calendar systems enable groups and organisations to maintain individual calendars and share these related to common events and resources (such as meetings and meeting rooms), while scheduling systems enable automated search through these calendars for finding available time slots for meetings. Scheduling systems are also often integrated with e-mail for automated invitations to meetings. These systems are also used to maintain shared calendars for common resources, such as meeting rooms and PC projectors, and coordinate their use.

The overall focus of time management is common for both calendar and scheduling systems, and many products combine this functionality. In addition to special products for these purposes, this functionality is also found in integrated products such as Lotus Notes, MS Exchange and Novell GroupWise. Knudsen and Wellington (1997) distinguish between different categories of these systems, aimed at supporting various levels in the organisation. Applications targeted at the workgroup level offer more support for individual tasks but are restricted on scalability. At the other end of the scale, enterprise level packages are focused on scalability but are more limited regarding personal productivity features.

There are obvious potential benefits from using this type of system, in the form of reduced time for negotiating and scheduling meeting times. Despite being available for more than a decade, however, this technology really has gained ground only recently. This is explained by improved network and client–server architectures, better user interfaces, improved support for individual tasks and better e-mail integration (Grudin and Poltrock, 1997). Calendaring tools are also bundled with hand-held devices such as palmtops/personal digital assistants (PDAs) and enable synchronisation of these calendars with common calendars maintained by a workgroup or organisation. The increasing use of such portable devices also contributes to the further diffusion of these systems. Another trend related to this type of technology is web enabling, making it possible for users to maintain their calendars through web browsers. Further, increasing inter-organisational collaboration (as in B2B e-Commerce) creates a need for integrating calendars to schedule meetings across organisations. However, this poses security challenges (Knudsen and Wellington, 1997). Another challenge is related to improving the interoperability between different calendaring and scheduling products. There are currently ongoing standardisation efforts for solving these issues by the Calendaring and Scheduling Working Group of the Internet Engineering Task Force (IETF, 2002).

Although calendar and scheduling systems now have become part of the everyday infrastructure in many organisations, there are still implementation challenges. Examples of this are users' preference for paper-based calendars (that are fully portable and can be "scribbled" upon), privacy (who should have access to view your calendar) and policies related to scheduling of meetings (who should have the right to book your time).

2.3.5 Integrated Products

This category covers products that incorporate functionality across other categories, typically some combination of communications, shared information space and coordination technologies. These can again be divided into three sub-categories: *collaboration product suites, integrated team support technologies* and *e-Learning technologies*. These include many of the same services, however, and are distinguished here more on the basis of the scope and focus of application. Typically, collaboration product suites comprise a comprehensive range of tools/applications, as well as serving as a network infrastructure. Some of these products also include an application development environment. Widespread examples of collaboration product suites are Lotus Notes/Domino and Microsoft Exchange/Outlook.

Integrated team support technologies are typically smaller in scale and scope than the collaboration product suites. A growing number of these are fully web-based, offering access to the services through web browsers. Finally, e-Learning systems are based on many of the same services as the two former sub-categories, but with the focus on pedagogical applications of these services in the form of course support environments rather than project support.

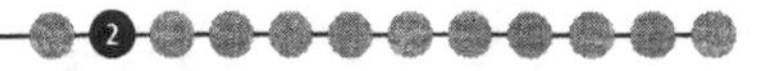

Collaboration Product Suites

The most widely known product in this category is Lotus Notes (or Notes for short) with the first version being launched in 1989. This product combines e-mail, calendaring and scheduling, threaded discussions, document management and workflow capability. In addition, Notes also contains an application development environment based on a scripting language. Thus Notes is often referred to as a platform for developing collaboration tools. With the release of the Domino server in 1997, Lotus put increasing focus on web-enabling Notes, including automated http translation of Notes databases and the possibility of accessing these through a standard web browser. With Release 5 (or R5) of Notes/Domino, the web integration has been further improved.

Through the replication functionality, databases at various sites can be synchronised at user specified frequencies, providing access to updated information for every user. This also makes it possible for mobile users to work with information locally at their workstation (e.g. laptop PCs) and replicate toward databases on the Notes server when convenient. The security features in Notes consist of a hierarchy of seven access modes, enabling detailed control of access to databases, documents, and even down to single fields in the documents. The main categories of applications that can be developed with Lotus Notes include broadcast applications (newsletters, industry bulletin boards, etc.), libraries of reference information (e.g. company policies, procedure manuals, software documentation), discussion databases, tracking applications (e.g. customer service tracking, project tracking) and workflow applications (processing of purchase requests, insurance claims, etc.).

Microsoft Exchange has also taken an increasing share of this market, covering much of the same functionality as Notes. However, the support for application development and integration of features in Notes is still more complete than for Exchange. To achieve the same with Exchange requires greater use of third-party extensions. Other products in this market are Novell Groupwise and Teamware, although these systems are not as widespread on a global scale as the two others.

The current development in so-called peer-to-peer (P2P) technologies represents an exciting new avenue for collaboration support. This technology allows direct, real-time connection between PCs, regardless of firewalls and company boundaries, without being dependent on centralised servers or databases. *Groove* is so far the P2P collaborative product that has gained most attention in the market. Developed by Ray Ozzie, the man behind Lotus Notes, this product offers a range of collaborative tools, including text and voice-based instant messaging, discussion lists, file sharing, project management, group calendar and co-editing of documents. The product also includes synchronisation services similar to the replication functionality in Notes, enabling offline use of the system. Further, Groove also offers a development environment for customisation and development of new solutions.

In general, P2P is a decentralised architecture, and companies looking to deploy this on a large scale face challenges related to network administration and security.

Integrated Team Support Technologies

This broad category comprises the growing number of products aiming to support various forms of teamwork by offering a teamwork environment with several asynchronous and synchronous services (see Appendix B for a list of example products). Products that focus on asynchronous services, such as document repositories, e-mail, discussion lists, and calendars, are often referred to as *teamrooms* or *projectrooms*. An increasing number of these also offer instant messaging/chat to support some synchronous interaction and awareness.

Products focusing on synchronous communication are often referred to as *desktop conferencing systems* or *real-time conferencing tools*. These systems offer integrated video, audio and data conferencing available from your PC, if equipped with a camera and audio equipment (microphone and loudspeakers). For example, these tools allow you to see a low resolution picture ("talking head") of the persons attending the conference, while viewing and annotating common information objects through application sharing or a shared whiteboard. However, the audio part in many of these systems does not yet equal telephone quality, and many organisations thus prefer to use separate audio/telephone conferencing in combination with the data conferencing functionality in these systems.

e-Learning Technologies

Use of IT for supporting instructional and learning processes is not a new phenomenon. Video, audiotapes and computer based training (CBT) programs have long been used as a substitution for classroom teaching, enabling individuals to follow courses and training programmes at their own leisure and speed. With the introduction of the CD-ROM, these learning devices have been augmented with multimedia features, making the learning experience more varied and captivating. Still, however, these tools are restricted to supporting one way interaction. The Internet has changed this by supporting the development of virtual learning environments with two-way interaction between students and instructor, as well as between fellow students. This has spurred the growth of a new industry termed the e-Learning industry. IDC projects e-Learning revenues in the corporate segment alone to grow from US$1 billion to $11.4 billion in 2003 (Ruttenbur et al., 2000). A survey of this market conducted by a US investment firm identifies three major categories of products that all are based on a combination of different collaboration technologies (ibid.):

- *Synchronous Shared Learning (SSL)* is based on same time interaction between learners and instructors, through a web-based collaborative learning environment. This can thus be thought of as a virtual classroom, where students participate at the same time but from different locations. Examples of technologies that may be included in this type of setup are one-way video, one- or two-way audio, application sharing and chat.
- *Asynchronous Shared Learning (ASL)* enables students to participate at different times, to better suit their schedules. Instruction is provided through the use of written material combined with video and audio clips, and there is no real time interaction with the instructor or class. Still, the students may interact with each other and the instructor through e-mail and bulletin boards enabling threaded discussion. These technologies are offered through a password-protected website.
- *Independent e-Learning (IEL)* is the term used for courses taken on an individual basis. Such courses are mostly offered by commercial vendors and let the learner complete different modules at his/her own pace. This type of learning environment will incorporate many of the same technologies as for ASL, although the interaction will mainly be between the system and the learner as there are no instructors or fellow students.

These product categories together support three different market sectors: the academic market, the corporate market and the consumer market. The general trend is towards more integrated learning systems offering learning on demand and highly customised courses. However, the further growth of this industry is believed to be dependent on developing standards that allow reusing content between courses from different vendors and interoperability of technologies. Some examples of e-Learning products are provided in Appendix B.

2.4 Current Trends

The area of collaboration technologies evolves fast, and any prediction on "the future of collaboration technology" quickly becomes obsolete. This final section will thus only summarise the key trends that already are influencing the further development of this area.

2.4.1 Integration of Services

As discussed in relation to the framework in Table 2.1, there is a general trend towards the integration of several collaboration features and functionality within a single product. Collaboration product suites such as Notes/Domino and Exchange/Outlook have so far been dominating in this area, but a growing number of web-based collaboration products are eating into this market. Further, we also see how collaboration features such as document management and workflow become integrated with other major organisational systems, such as ERP systems. Typically, the boundaries between these product categories tend to blur.

In general, it is likely that the term collaboration technology will gradually cease to exist as a separate "phenomenon", and will instead be incorporated as a natural feature in office support tools and organisational IT systems. This can already be seen to happen in the areas of B2B e-Commerce, e-Learning and knowledge management, where the underlying infrastructure of collaboration technology is presumed to exist as a natural part of these technologies. With this development many of the barriers to effective use of collaboration technologies will also be eliminated, such as interoperability and users having to switch between separate products. The general improvement in network services is also an important factor, making bandwidth constraints more and more obsolete.

2.4.2 Web-based Collaborative Tools

The move towards the Internet/web as the platform for collaboration technologies has been a common theme in the description of the different technologies throughout this chapter. This can be seen to take two forms: web-enabling of existing products supporting access through web browsers running on thin clients (such as for Notes/Domino), and development of new web-based products challenging the existing products. The first of these trends has already come a long way and can now be regarded more as a requirement than a feature offering any competitive advantage. However, the development of "pure" web-based products is only in its infancy, and it is expected that the coming years will bring many new and exciting products.

2.4.3 Mobility of Services

Lack of portability has been a general barrier to effective deployment of collaboration technologies. Examples of this are the use of calendar systems, meeting support systems etc. Although laptops combined with mobile phones today already enable anytime/anyplace connectivity, the development in mobile computing and wireless services represents more radical improvements. For example, being able to bring your laptop computer to every meeting and connect to any intranet or Internet service may enhance the meeting productivity and quality of outcome. The rapid expansion of services integrated in hand-held devices such as palmtops/PDAs also contributes to an increasing mobility of services.

2.4.4 Advanced Multimedia

Most electronic communication among dispersed groups today is still asynchronous and text-based, although desktop conferencing is gaining ground. The value of static "talking heads" in today's desktop conferencing systems is questioned, as this cannot mediate body language. In general, finding ways of creating shared awareness among the participants in virtual collaboration spaces is considered an important challenge in the further development of these services. Current research in different media labs is focusing on how to deploy advanced 3D models and virtual reality (VR) to create virtual representations of participants that can convey body language, gazes, movement, etc. (Luciano et al., 2001). Another important application area is collaborative virtual design environments, currently used in aircraft and car design, for example (Ragusa and Bochenek, 2001; Smith, 2001). Here, VR is applied to product and system design activities, allowing the viewing and review of entire systems, assemblies, and parts. Similarly, 3D modelling is also being used to develop new interfaces for document repositories, enabling the user to navigate in landscapes of document objects. In general, we are only at the beginning of utilising the potential of these technologies for enhancing existing services.

2.4.5 Boundary-spanning Applications

Another general trend is the use of collaboration technology outside the boundaries of the "traditional" organisation. Different forms of virtual teamwork involving participants from several organisations have long been taking place in industry, but with the current focus on B2B e-Commerce, value chain integration and so on, this is becoming far more extensive. While IT support for inter-organisational collaboration previously was limited to establishing static channels for data exchange (e.g. EDI), the focus is now more on also supporting interpersonal interaction, knowledge sharing and learning among the participants, as exemplified by the notion of c-Commerce discussed by the Gartner Group. New products based on peer-to-peer technology here also offer new possibilities for effective cross-organisational collaboration.

Further, there is also increasing use of collaboration technologies on a societal level, as in online communities, teledemocracy and e-government. As different electronic channels such as instant messaging and desktop conferencing become a natural way of communicating, this opens new and interesting possibilities for increased interaction with the community in public decision processes.

3

Implementation of Collaboration Technologies: A Review of the Literature

Bjørn Erik Munkvold

3.1 Introduction

This chapter presents some findings and experiences from previous research related to the organisational implementation of collaboration technologies in industry. The studies included are mainly collected from academic publications, to ensure that the findings are based on an "objective" stance and a thorough methodological approach. One of the aims of this book is also to contribute to disseminating the findings and practical implications from academic research to a broader audience of industry practitioners. However, vendors and consulting companies have also generated much useful material in the form of industry and technology reports, and "white papers" presenting exemplar case studies. Although not included in this review, these sources may offer a useful complement to academic research, as long as one is aware of the potential bias and/or sales aims in these sources.

The chapter starts with defining some key concepts related to the implementation process. This is followed by a brief review of a classic work (Grudin, 1994b) discussing the challenges of development and implementation of collaboration technology. The rest of the chapter presents selected examples of field studies of the implementation of different collaboration technologies according to the classification framework presented in Chapter 2 (Table 2.1). Thus, the chapter also provides a brief status report of the implementation research for each of these technologies.

A note to the reader may be relevant here. Chapters 3 and 4 together present a summary of previous research on implementation of collaboration technologies that may serve readers with various levels of interest in this area. For each category of technology covered in this review, a brief summary is provided. Chapter 4 summarises the implementation factors according to a simple taxonomy. A guide for the different reader groups is thus:

- Readers only interested in implementation factors without research background, skip to Chapter 4.
- Readers interested in a brief summary of implementation research for different collaboration technologies, read summaries in Chapter 3 plus Chapter 4.
- Readers interested in implementation research for a specific technology, read the related section in Chapter 3.
- Readers interested in research on implementation of collaboration technology in general may read both chapters in full.

3.2 Some Definitions Related to Implementation

The term *implementation* is used differently in different communities. In areas such as computer science, human–computer interaction (HCI) and software engineering, this term basically refers to the actual coding of the system, while in information systems (IS) research and practice the term denotes the process of introducing the technology in an organisational setting. Grudin (1993) discusses how differences in terminology may constitute a barrier to effective communication between these communities. This section provides definitions of some key concepts as they are used in this book.

The focus in this book is on implementation according to the IS perspective, also referred to as *organisational implementation* (Walsham, 1993). There are also several related terms that require clarification. Building on a diffusion of innovation perspective (Rogers, 1983), Kwon and Zmud (1987) introduced a six stage model for the IS implementation process, covering all stages from project initiation and acquisition of a new technology (through purchase or in-house development) to the stage where the technology is fully "internalised" in the daily work practices and full benefits from the technology may be realised. Table 3.1 briefly summarises these stages.

In this model, the term *adoption* is used at the organisational level, referring to "the decision to make full use of an innovation" (Rogers, 1983). The adoption by individual users, referred to as *acceptance* in the model, thus takes place after *adaptation* of the technology in the organisation through the activities listed in Table 3.1. According to this perspective organisational adoption does not "automatically" imply adoption by the individual users. In other words, an organisation may decide to invest

Table 3.1 Six stage model of the IS implementation process (adapted from Cooper and Zmud, 1990)

Stages	Activities
Initiation	Scanning of organisational needs and IT solutions.
Adoption	Negotiations to get organisational backing for IT implementation.
Adaptation	Developing, installing and maintaining the IT application. Developing new organisational procedures. Training of users.
Acceptance	Inducing the organisational members to use the technology.
Routinisation/Use	Use of IT application is encouraged as a normal activity.
Infusion	The intended benefits from the technology is obtained through effective use of the technology.

in a technology and make it available for use throughout the organisation, while the individual users for various reasons may still decide not to adopt the technology. The literature review presented in this chapter includes several examples of this.

This model can be useful for classifying factors influencing the different stages of the implementation process (see Munkvold, 1998a for an example of this). Despite the "linear" presentation of stages in the model, however, the implementation process should rather be viewed as iterative, and with the different stages partly overlapping. For example, it could be argued that the process related to user acceptance actually starts early in the process, being interrelated with adoption and adaptation.

Further, in more "user centric" communities such as HCI, adoption is only used to mean an individual user's decision to use a technology. Accordingly, in these communities the activities related to adoption and adaptation in Table 3.1 are referred to as *deployment* of the technology, in the sense of making it available for use (see also Chapter 8 for a more in-depth discussion of these terms from a HCI perspective).

Another term frequently used related to the implementation process is *diffusion*, defined as "the process by which an innovation is communicated through certain channels over time among the members of a social system" (Rogers, 1983). In the context of collaboration technologies, this refers to the process by which the adoption and use of the technology spreads throughout an organisation, both as a result of planned distribution as well as emerging social mechanisms such as peer pressure. Finally, the term *assimilation* is sometimes used for describing how the technology is "taken up" by a user community. This corresponds to the last three stages in the model in Table 3.1 (acceptance, routinisation and infusion).

Based on the author's perspective, the IS terminology is used throughout most of this book. To avoid confusion related to the adoption term, this is qualified either as organisational adoption or user adoption. The case chapters in Part 2 of the book to some extent reflect the invited authors' preferences regarding implementation terminology. The definitions presented in this section may thus serve as a simple thesaurus, outlining how the different terms are related.

3.3 Challenges in Development and Implementation of Collaboration Technologies

One of the first works to address the problems inherent in design and evaluation of collaboration technology was an article entitled "Why groupware applications fail: problems in design and evaluation", presented by Jonathan Grudin at the second CSCW conference in 1988 (later published as Grudin, 1989). The article discusses different types of problems in groupware design and evaluation as compared to single-user applications and mainframe systems, with examples from different collaboration technologies such as automatic meeting scheduling, voice mail and e-mail. The arguments were developed further in Grudin (1994b) where he presents eight challenges in groupware development and implementation, listed in Table 3.2:

Table 3.2 Eight challenges for groupware developers (Grudin, 1994b) © [1994] ACM, Inc. Reprinted by permission.

1. *Disparity in work and benefit.* Groupware applications often require additional work from individuals who do not perceive direct benefit from the use of the application.
2. *Critical mass and "prisoner's dilemma" problems.* Groupware may not enlist the "critical mass" of users required to be useful, or can fail because it is never to any one individual's advantage to use it.
3. *Disruption of social processes.* Groupware can lead to activity that violates social taboos, threatens existing political structures, or otherwise demotivates users who are critical to its success.
4. *Exception handling.* Groupware may not accommodate the wide range of exception handling and improvisation that characterises much group activity.
5. *Unobtrusive accessibility.* Features that support group processes are used relatively infrequently, requiring unobtrusive accessibility and integration with more heavily used features.
6. *Difficulty of evaluation.* The almost insurmountable obstacles to meaningful, generalisable analysis and evaluation of groupware prevent us from learning from experience.
7. *Failure of intuition.* Intuitions in product development environments are especially poor for multi-user applications, resulting in bad management decisions and an error-prone design process.
8. *The adoption process.* Groupware requires more careful implementation (introduction) in the workplace than product developers have confronted.

Grudin explains how the first five of these challenges require better knowledge of the workplace of the intended users, while the final three require changes in the development process. Especially the issue of perceived *disparity in work and benefit* among different stakeholders in collaboration technology implementation has gained widespread attention from researchers in this area. In many cases, those designing and/or implementing collaboration technology are not attuned to the fact that the technology might not be perceived as equally beneficial among all stakeholder groups required to adopt the technology, due to the extra work created from recording and maintaining information, etc. As will be illustrated in several cases in this review, this extra overhead may actually constitute a serious threat to an implementation project.

Another key concept related to implementation of collaboration technologies is *critical mass*. Rooted in economy and sociology, the theory of critical mass explicates the conditions under which reciprocal behaviour gets started and becomes self-sustaining (Markus, 1987). When applied in the context of communications media this term refers to the number of users that have to adopt the technology before the adoption of the technology becomes self-sustaining. Establishing a critical mass of users requires *universal access* to the communications medium, that is, the ability to reach all members through the medium. If universal access is not achieved, this will reduce the benefits for the individual users. In this way, communications media are different from other technological innovations in that the interdependence between early and later adopters is reciprocal rather than sequential. The benefits for early adopters of these technologies are less than for later adopters. If a critical mass of users is not established, the early adopters may discontinue use.

This illustrates the *interdependency* inherent in user adoption of many collaborative IT applications. Unlike single-user tools, the adoption of many collaboration technologies is actually interdependent among the users in that the benefits and costs of the applications to one user are contingent on the behaviour of other users. Markus and Connolly (1990) discuss how this interdependence can lead to failure in implementation of these technologies.

One example is the implementation of a common database, like an automatic meeting scheduling system, that is based on the contribution from each user in the form of input to the system. In this type of system, the benefits for both adopters and non-adopters increase as the number of adopters increase. However, since the information in the database is available for both adopters and non-adopters, non-adopters will always be better off than adopters because they do not have to pay the contribution costs associated with inputting the data. In this way, the self-interests of the individual users may actually result in behaviour that defeats the interests of all users, such as "free-riding". Some form of managerial intervention may then be necessary for ensuring that the applications are used.

At the core of Grudin's argument is that those responsible for development of collaboration technologies need to be sensitive to the additional problems and complexity that arise when developing software for supporting groups, as compared to single-user applications or large enterprise systems. Based on the original focus of groupware as technology supporting small groups, Grudin also argues that this technology will often fail to receive the same attention and management backing as "traditional", organisation-wide systems such as accounting or payroll systems. Today, however, there is also an increasing number of organisation-wide, collaborative applications based on technologies such as Lotus Notes and intranets, that have high visibility in the organisation. Nevertheless, the challenges identified by Grudin still prove relevant in several of the implementation projects presented in this book.

3.4 Empirical Findings from Previous Field Studies

This section presents some findings from previous field studies of the implementation of different collaboration technologies in various industrial settings. Following the classification framework introduced in Table 2.1, the current status regarding implementation research for each type of technology is briefly presented, together with practical findings related to this technology. Each section concludes with a brief summary of the implementation research for this type of technology.

3.4.1 Implementation of Communication Technologies

This category includes both asynchronous and synchronous communication technologies. This section presents findings from field-based research on the implementation of the two most dominating communication technologies so far, e-mail and videoconferencing. A recent study on the use of instant messaging documents how this technology enables new forms of communication and interaction in the workplace. In addition, findings from implementation studies of some integrated communication "packages" are briefly presented, illustrating the importance of establishing a critical mass of users and strategies for obtaining this.

Organisational Implementation of E-mail

E-mail is often referred to as the most successful collaboration technology regarding diffusion and user adoption. This is explained by the intuitive nature of this service and the clear analogy to "traditional mail" (Bullen and Bennett, 1990). However, while many research studies have focused on the organisational effects of introducing e-mail (see for example Finholt and Sproull, 1990; Kock, 1999; Sproull and Kiesler, 1986) or the nature of this communication (Sproull and Kiesler, 1991; Kock, 1999), fewer studies have focused explicitly on the process of implementing this technology. A possible explanation for this may be that the relatively simple functionality of e-mail has attracted less attention from the implementation research community compared to more complex collaboration technologies.

Despite the simple and intuitive functionality, however, there may still be barriers to effective use of this technology. Owing to the general familiarity with e-mail among today's employees, user acceptance does not represent a problem in the implementation of this technology. As a result, e-mail is often implemented without much emphasis on training or practical guidelines. This may actually result in problems related to ineffective use of the technology. For example, a survey of e-mail users in a Norwegian public service organisation half a year after implementation, showed that 50 per cent of the users still had problems with the system and did not know how to take advantage of all its features (Kautz, 1996). Lack of guidelines may result in infrequent use (such as not checking the mail box daily) and uncritical distribution of documents as e-mail attachments to a large number of employees without considering who really needs this information.

For distributed organisations that rely heavily on electronic communication, establishing protocols for effective communication is especially important. This may include creating closed "communication loops" by always verifying the receipt of messages, practising "active electronic listening" through e-mail, and assigning appropriate priority to messages. In addition comes the development of general norms for what is to be considered acceptable "tone" and forms of electronic communication in the organisation, sometimes referred to as "netiquette". For example, the Internet Engineering Task Force (IETF) has suggested extensive netiquette guidelines (IETF, 1995).

Implementation of Videoconferencing

The diffusion of this technology in industry has been slower than originally predicted. Costly investments, technological requirements and a relatively high learning curve have acted as barriers (Egido, 1988; Sanderson, 1992). Further, as for all synchronous collaboration technologies, time zones may constitute a barrier for use of this technology in global companies and projects. Research on the organisational use of videoconferencing and the related effects has also been fairly heterogeneous regarding focus, methodological approaches, research design and equipment used, with findings being equally mixed (Finn et al., 1997).

With the development of PC-based desktop videoconferencing systems several of these implementation problems have been eliminated, and the rapid development in these products offers improved quality and accessibility of services for less costs. An increasing number of companies thus see this technology as an important means for reducing travel costs, and recent world events such as the September 11 World Trade Centre bombings contribute to a trend where companies look for alternatives to extensive business travel.

Perey (1997) presents some advice for assessing whether a company is ready to implement this technology. Among the indicators suggested are:

- a well-established IS department, preferably with previous experience from deployment and use of collaboration technologies;
- up-to-date infrastructure including PCs with multimedia, a well-functioning corporate network and high-capacity servers;
- incentive systems for rewarding employees who show initiative and good results in the use of videoconferencing.

Further, she recommends conducting detailed "ethnographic" user surveys for being able to map important factors that may influence the user acceptance of this technology. These factors include frequency and patterns of travel, whether performance measurements in groups are related to real-time interaction or not, and the users' past experience with the technology. Based on these surveys, corrective measures may be initiated that can increase the users' receptivity of the new technology. Other types of surveys include project surveys for identifying projects with a high collaborative potential (e.g., involving synchronous interaction and/or tight deadlines), and manager surveys to identify potential champions who may lead the company's utilisation of the technology further.

Instant Messaging in Action

One of the first research studies on organisational use of instant messaging (IM) is presented by Nardi et al. (2000). They studied the IM usage of 20 people in a telecommunications and an Internet company, using a combination of interviews, observation and logs of IM sessions. They found that the flexible and immediate nature of IM supported a range of informal communication tasks: quick questions and clarifications, coordination and scheduling, organisation of impromptu social meetings, and keeping in touch with friends and family. Compared to e-mail, the immediacy of IM makes it more suitable for scheduling, while also avoiding the more lengthy interaction of a phone call. The immediate give and take of IM conversation was also often found to be more informal than e-mail, with frequent instances of jokes and humour.

By providing awareness information about the presence of communication partners, IM also served an important function in negotiating the availability of others to initiate a conversation. When a session was initiated IM was also used to manage the conversational process, including switching to other communication media (phone, e-mail, etc.) during the interaction. This type of media switching coordinated by IM can be expected to become increasingly widespread, as IM systems become integrated with audio, video and data conferencing. The flexible interaction forms enabled by this thus calls for guidelines for effective media switching and use.

Critical Mass in Distributed Organisations

In implementation projects spanning organisational boundaries, the process of establishing a critical mass of adopters may be more challenging than for intra-organisational implementation. An example of this is a study of an organisational network of small and medium sized enterprises (SMEs) in the building construction industry (Munkvold, 1998b). As part of their strategy of becoming a virtual engineering company, they implemented an ISDN-based communications package offering integrated e-mail, fax, telephony and file sharing for supporting an electronic tendering process. Several barriers to effective implementation and use of the technology were observed in this project:

- Various delays in the project resulted in a time lapse of three months between the first and last of the companies getting connected to the collaborative infrastructure. At the end of this period, the early adopters had more or less abandoned the system due to lack of universal access to the system, and a critical mass of users was never reached.
- Network incompatibility created technical problems in the early stages of the project, resulting in the users developing distrust in the technology. Some users thus started to send fax messages as "backup" to the e-mail "to be sure the message got through". Once this distrust had developed, it was hard for the implementers to rebuild trust in the system.
- Low computer experience in some of the companies also resulted in ineffective routines. For example, the manager in one of the companies had the secretary handle all e-mail on his behalf.

Chapter 6 discusses similar challenges faced by another company, Kværner, in establishing a critical mass of users for their global information network.

Table 3.3 Implementation strategies suggested in Ehrlich (1987b)

1. Clearly identifying communication problems
2. Effectively matching computer system solutions to existing problems
3. Avoiding pilots with inappropriate groups, i.e. pilots should have demonstrated needs for communication because of distance, completeness or other factors
4. Understanding expert/mature use of system features, i.e. observing mature system use to help in forming a strategic direction for the new system, and identifying the potential of the technology. This can be obtained by visiting mature system users at other sites
5. Providing education that demonstrates positive impact on the work day
6. Providing step-by-step training on unfamiliar features
7. Encourage top management use of system features
8. Providing follow-through support to encourage system use
9. Troubleshooting systems problems quickly to avoid premature rejection

Experiences from the Implementation of Office Communication Systems

Ehrlich (1987b) conducted eight studies of the implementation of office communication systems (e-mail, electronic calendars and answering machines) in three organisations, based on questionnaires and user interviews. Although the technology addressed many of the communication problems in the organisations, there was still initial resistance at some sites. Ehrlich thus argues that careful strategic planning is required in order to ensure successful implementation of this type of technology, and describes nine implementation strategies that were found to be important in the cases studied (Table 3.3).

As argued by Ehrlich, these implementation strategies may be considered relevant beyond the technologies addressed in her study. This is confirmed by the experiences from the implementation projects presented in Part 2 of this book.

3.4.2 Implementation of Shared Information Space Technologies

Shared information space technologies include document management systems, data conferencing and electronic bulletin boards. Few studies have been identified in the collaboration technology research literature that focus explicitly on the organisational implementation of document management systems and bulletin boards. However, related to the focus on knowledge management (KM), these technologies are addressed as part of the implementation of a "KM infrastructure". There is also a growing focus on content management as an overarching information management strategy in organisations, but there is still little field-based

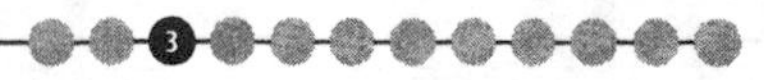

Brief Summary: Research on Implementation of Communication Technologies

Field-based research on communication technologies has focused more on how electronic media affect communication and interaction than on the process of organisational implementation of these technologies. Establishing universal access to the services and building a critical mass of users are key issues in the implementation. E-mail has so far been the most widespread communication technology, but instant messaging is increasingly being used for more informal and immediate communication and coordination. Technical limitations in the form of bandwidth and image quality have previously been a barrier to the widespread diffusion of videoconferencing, but desktop videoconferencing is now gaining ground. A major challenge related to communication technologies is to establish guidelines and norms for effective use.

research on the implementation of this type of solution. As for data conferencing, studies of organisational adoption and use of this relatively new technology are starting to appear.

Challenges in the Transition from Paper-based to Electronic Document Production

In an article entitled "The Work to Make a Network Work", Bowers (1994) reports the findings from a two-year field study of the implementation of a local area network for running CSCW applications (e-mail, meeting scheduling, document annotation and shared authoring, etc.) in an organisation within the UK government. Based on user interviews and field observation, he studied the process of installation and use of the CSCW network. The study describes how the network ran into several difficulties, and was resisted by its potential users for various reasons. Although the study covers several collaboration technologies, the most interesting findings are related to the transition from paper-based to electronic document production.

The introduction of new technical possibilities was found to provoke debates about existing practices. An important question in the introduction of the technology became whether the existing documents and related practices should be changed, or if the technology should be rejected as being inconsistent with the existing practices. For example, in addition to making the documents easily available the network also made the practices used for producing the documents "visible, inspectable and manageable". This led to suggestions of changing the temporality of existing work practices, such as using "continuous" status reporting instead of monthly reports. However, this raised several questions related to who should have access to these status reports, and at what stage in the process "work-in-progress" documents should be made available in the network. These issues actually resulted in much of the "real work" of the document production remaining off the network.

As senior civil servants often work at remote locations without access to computer technology, the limited portability of electronic documents for them resulted in a preference for paper copies of the documents. It also proved difficult to establish fully automated routines for document production, due to problems with distributing documents to persons outside the network. Partial solutions here were not satisfactory, due to the extra work needed for translating the documents between electronic and paper-based form. This could thus result in these facilities being restricted altogether.

The ability to share documents was also found to result in dilemmas of responsibility and ownership. Further, the users expressed uncertainty about the degree of formality to be associated with electronic documents. Usually, it was perceived as somewhere between speech and writing.

Some users felt discomfort with the accountability resulting from contributing to bulletin boards and computer conferencing technologies, and thus were reluctant to use this technology. Several users did not trust the network. For example, some early delivery failures made one of the users print out copies of all e-mail sent and received.

Bowers concludes by stating that the complexity of collaboration technologies may lead to an even greater management overhead than with more conventional tools. For example, network implementation, management and use requires the solution of a range of problems, thus being termed as "heterogeneous engineering". The different manifestations of extra work needed when implementing collaboration technologies are often regarded as trivial when viewed individually, but taken together "the work to make the network work" may actually be a threat to the viability of the implementation project.

Implementation of Knowledge Repositories

Davenport and Prusak (1998) report on industry practices related to the conduct of knowledge management projects. Based on studies of more than 30 leading companies worldwide, they found that these projects usually comprise one or more of the following three elements:

- creation of knowledge databases or knowledge repositories;
- improved access to knowledge in the company ("who knows what") and the transfer of this knowledge;
- improved "knowledge culture" and "knowledge environment"

These projects include different technological architectures. The knowledge databases are based on technologies such as Lotus Notes databases or document management systems. A general challenge related to the development of these systems is to establish a set of common keywords for information searches. Several examples exist of companies that have not been able to fully utilise their knowledge/experience databases due to complex search procedures. One example here is the Knowledge Xchange system in Accenture (former Andersen Consulting). Employees have here experienced a learning curve of 5–7 years (!) before gaining sufficient overview and in-depth expertise to be able to use the databases effectively by "knowing where and what to look for". For this type of company-wide databases, a thesaurus of indexes is necessary for making the knowledge searchable for the different parts of the company. In addition, clear roles and responsibilities need to be established for maintenance and quality assurance of the database contents.

Implementation of Data Conferencing

Data conferencing combines text messages (chat), shared whiteboard and application sharing. There are not many field studies on organisational implementation of this technology, as the use of this in organisations is only just starting to grow. In the following, some experiences from the adoption and use of this technology in three engineering companies in different industries are briefly presented.

A field trial of NetMeeting in Bell Labs

Finholt et al. (1999) present a field trial of use of Microsoft NetMeeting to support collaboration between teams of software engineers in Bell Labs, located in England and Germany. Although the interviewed users perceived the features of the tool to be very useful, and expressed "moderate satisfaction" with the product, the use during the three-month trial period was low. This was explained by a lengthy installation period (NT users not having administrator privileges allowing them to install the software themselves) allowing little time for real use during the trial period, and one of the teams using UNIX machines that did not run NetMeeting. Only the one team having a "very enthusiastic NetMeeting supporter" reported more frequent sessions.

Some weaknesses and technical problems were also reported, such as slow refresh ("motion sickness") when moving or resizing shared windows, and the graying out of windows of non-shared applications blocking shared windows. Further, users reported periodic difficulty in accessing NetMeeting servers, and also experienced problems with locating other users due to the lack of a common default server. Finally, there were also some problems due to incompatibility between different versions of NetMeeting. To overcome problems with confusion related to sharing application control, parallel phone conferences were initiated and a mediator was often appointed to "orchestrate transfers of control".

Implementation of NetMeeting in a Norwegian engineering company

Similar findings were reported in a longitudinal study of adoption and use of NetMeeting in a Norwegian engineering company (Line, 1998). As part of their increased focus on distributed collaboration, the company ran a pilot test of NetMeeting involving 12 users at different locations. Use patterns were found to develop slowly, and it took a long time for the users to incorporate this tool into their workday. This can also be explained by the nature of the implementation as "natural dissemination", without any formal implementation project. Despite being conceptually simple, the service still had an "aura of advanced technology" around it, imposing a threshold to initiate its use. Two different types of use were observed:

- technical support and coaching, involving configuration of remote computers by the users themselves supported by a NetMeeting session, and support in the use of a 3D terrain modelling application;
- "open cooperation", involving support for tasks such as the pricing of tenders, discussion of project schedules or report outlining. This commonly comprised 15–30 minutes sessions, sharing Microsoft Office, project planning and CAD applications.

After a period of about eight months more advanced use of the service started to emerge, such as using NetMeeting for long sessions directly supporting physical meetings also involving external partners (e.g. related to bid preparation). In general, two modes of connections were prevalent. These were "by appointment" - that is, scheduled in advance by phone or e-mail - and ad hoc during a telephone call. Frequent users of NetMeeting mostly reported the latter form, having included NetMeeting in their repertoire of work processes. Due to the latency involved when applying NetMeeting for supporting audio, a parallel telephone meeting was normally used. Videoconferences were also tested but not evaluated in actual meetings. Comments from the users during these tests indicated that video was regarded more as an "interesting toy" rather than a useful communication tool.

Experiences from Boeing

The Boeing company makes extensive use of data conferencing for supporting distributed meetings. Based on observation of four permanent teams, Mark et al. (1999) describe the practices developed around the use of NetMeeting in the Boeing company. In contrast to the other two cases presented above, user adoption of the technology was relatively fast due to the increasing use of distributed meetings. Still, several barriers were identified for making effective use of the technology.

Without an assigned role of "technology driver", establishing the NetMeeting sessions often took 10 minutes or more, representing a substantial loss of time for the 10–20 participants present. Further, problems were also observed due to different configurations being used at each site, and limited use of some of the functionality. During the meetings there were also frequent problems with coordinating interaction, involving difficulty in knowing who were present at each site and uncertainty about turn-taking during the discussions. Many people also conducted "multitasking" when attending these distributed meetings, such as reading e-mail or talking with other people in the room. While the participants considered this to be an advantage, it also lowered their involvement and commitment in the actual meeting.

As a result of these problems, some of the teams created new facilitator roles similar to ones used in face-to-face meetings. The technology facilitator would be responsible for all aspects of technology use, such as establishing a connection, trouble-shooting and controlling the presentation (e.g. through gesturing with the cursor and zooming in on relevant content). The virtual meeting facilitator took some of the load off the meeting leader, focusing on integrating the remote sites in the meeting discussion. This involved identifying who was speaking, explaining comments for the benefit of remote sites and probing their responses, and facilitating turn-taking in speaking.

One team also augmented their meetings with the use of NetMeeting's chat functionality for side conversations and communicating different types of information (solving problems, background data, etc.) without disrupting the meeting. Chapter 7 presents a more detailed account of the experiences related to the implementation of data conferencing in Boeing.

These three case examples illustrate that while companies are making increasing use of data conferencing, there are some thresholds that need to be overcome to develop advanced, routinised use of this service. Some common findings related to the use of this technology are:

- infrastructure requirements may constitute a barrier to adoption;
- the audio and video functionality were not used as part of this service, but instead separate audio conferences were frequently used;
- there is a need for a facilitator role for making effective use of the technology.

Implementation of a Multimedia Annotation Tool

Francik et al. (1991) report the experiences from the implementation of the Freestyle multimedia annotation tool at 15 sites. This system provided multimedia communication for networked PCs, with the possibility for exchanging computer screens or scanned-in papers with handwritten and/or voice annotations. Despite its simple design, inertia was experienced in the implementation of the technology. Three major issues were identified in this process. First, the customers had difficulty in identifying how the technology could be used in familiar work processes. There was a tendency to focus on individuals' use of the system, without considering support for entire work groups. Thus the justification for purchasing the technology for entire workgroups was not established. The system was also often used merely to speed up the form-based paper flow, rather than making use of the multimedia annotation features.

The second issue was related to the selection of workgroups for conducting pilot trials. As experienced in the Ehrlich (1987b) study, successful adoption of the system requires identifying members of a workgroup with a joint need for communicating. However, several factors were identified that influenced the selection of workgroups, resulting in pilot groups that were inadequate for the test:

- focusing on managers instead of a functioning workgroup;
- limiting the number of members in the pilot groups to minimise the resource requirements, thus excluding members that were involved in the business process to be supported;
- restricting geographic distribution, thus reducing the visible impact of the system;
- following organisation chart boundaries, thus excluding key communicators

The third issue identified was related to the assignment of equipment for using the system. This assignment was often made on the basis of organisational rank rather than function, thus resulting in low-ranking organisational members receiving inadequate equipment for carrying out their work tasks, which created bottlenecks in the work process. Key to solving these problems is that the vendor understands the work process implications of their designs, and works closely with the customers in developing strategies for overcoming the barriers described above.

3.4.3 Implementation of electronic meeting systems (EMS)

Most of the EMS-related research has been conducted in academic, "laboratory" settings, studying the impact of EMS support on process and out-

Brief Summary: Research on Implementation of Shared Information Space Technologies

The transition from paper based to electronic document handling introduces new challenges related to accessibility, ownership and maintenance of the information. Few studies in the CSCW research literature have focused explicitly on the organisational implementation of document management systems, but there is increasing focus on "content management" and technologies supporting the entire information lifecycle in the organisation.

Several field studies focus on the implementation of shared databases and knowledge repositories, for example Lotus Notes (see Section 3.4.5). These studies show that effective use of these technologies may depend on new roles and routines for information sharing, possibly also backed by explicit incentives. Effective search engines and clear responsibilities for quality assurance of database contents, are also important factors.

Data conferencing/application sharing is increasingly used for supporting distributed collaboration, in the form of distributed presentations and user training and support. Use of the technology for collaborative authoring is still less widespread. Technical challenges and the need for developing a "social protocol" regulating turn-taking and screen management are potential inhibitors for adoption. To overcome these barriers, several organisations use dedicated facilitators for technical support and integrating the different sites in the meeting process.

come for student groups working on assigned tasks and comparing this with "traditional" face-to-face teamwork. Extensive reviews of this research conclude that few clear effects have emerged from this stream of research (Fjermestad and Hiltz, 1999). However, a recent meta-study suggests a model for interpreting the findings on EMS effects on performance, with task fit and appropriation support (training, facilitation, etc.) as the main factors affecting performance (Dennis et al., 2001). When classifying the findings according to this model, fit between task and EMS tools selected and appropriation support were found to result in more ideas, less time used and more satisfied participants than for groups not supported by EMS. Yet, the lack of organisational context in these experimental studies clearly limits the possibility for generalising the findings from this research to use of this technology in real organisational settings.

With increasing proliferation of this technology in organisations, the number of field studies is also increasing. A recent review (Fjermestad and Hiltz, 2001) found more than 50 such studies, focusing on use of EMS in a range of different areas such as strategic planning, business process modelling, and requirements analysis and design. In general the results from these field studies are much more positive regarding the effects of the technology on the process and outcome of organisational meetings. Some examples of findings include (DeSanctis et al., 1993; Nunamaker and Briggs, 1997; Post, 1992; Tyran and Dennis, 1992):

- group productivity gains from EMS use – such as 50 per cent reductions in labour costs and 90 per cent reductions in elapsed project time – were found in studies at IBM and Boeing;
- broader and more active participation has been widely substantiated in both lab and field research;
- anonymity embedded in the EMS helps encourage more objective and constructive evaluation that improves the quality of ideas generated;
- buy-in and ownership of the meeting results is often increased by EMS use;
- EMS use may lead to improved decision quality, through increasing the number of creative ideas put forth, and stimulating more thorough problem analysis.

Various factors have been found to influence the EMS effectiveness. Active facilitation is found to constitute a key factor in electronic meeting success (Anson et al., 1995; Bikson and Eveland, 1996; Dennis et al., 2001). Task-technology fit is likewise a critical success factor (Dennis et al., 2001). For example, in several studies divergent tasks have been found to benefit from EMS use more than convergent tasks (Bikson and Eveland, 1996). In general, electronic media are used more to generate ideas, while verbal and mixed channels are used to discuss issues (Tyran and Dennis, 1992).

Despite an increasing number of field studies, there are still few studies that focus explicitly on organisational adoption of EMS. The unit of analysis is mostly at the team level, with most studies focusing on the appropriation of the technology by permanent teams. Further, much of this research is based on organisations that use these technologies at third-party sites. The costs involved in installing technologies like GroupSystems imply that many organisations find it too expensive to purchase a company licence, and instead prefer to use the technology through consulting services provided by academic institutions. Thus, there is limited availability of field studies addressing the organisational adoption and diffusion of EMS technologies. In the following, three studies focusing explicitly on the organisational implementation of EMS are presented in more detail.

Implementation of GroupSystems at the World Bank

Bikson and Eveland (1996) present a socio-technical analysis of the successful implementation of GroupSystems for supporting group decision-making in the World Bank. The implementation process can be described as a stepwise process lasting over a period of 5 years, and involving mutual learning between technologists and organisational developers. Several features of the implementation process are stated to have contributed to its success:

- a *high level champion* from the organisational behaviour department (ORG), resulting in a focus on the technology as a tool in organisational development;
- a close collaboration between the ORG department and the IT department, resulting in an "*unusual degree of socio-technical balance*";
- a *real user design/implementation team* comprising 12 carefully chosen members that were to become facilitators and technographers;
- a focus on *learning and training* throughout the implementation;
- *effective use of pilot trials*, allowing for experimentation with the technology.

All of these issues are well known from the innovation literature. Yet, as argued by the authors, it is rare to see these success factors being fulfilled to the same degree as in this case. However, the study also shows how the implementation process resulted in several organisational outcomes that were not planned at the outset. The authors see this as an example of the socio-technical dynamics, involving a mutual adaptation between the social and technical systems.

Implementation at IBM

Grohowski et al. (1990) describe one of the earliest major organisational adoptions of EMS, in IBM. Over a 3-year period, GroupSystems spread from one single site to 33 sites, used by over 15,000 participants throughout the company. By tracking the results of a sampling of meetings they substantiated very positive results. Based on these experiences they identified a set of success factors for organisational implementation, examples of which are organisational commitment, the need for an executive sponsor, training, facilitation support, dedicated facilities, cost/benefit analysis and meeting managerial expectations.

Implementation at Boeing

Post (1992) reports from the design and results of the evaluation study conducted by Boeing prior to their decision to purchase TeamFocus (the predecessor of GroupSystems). This involved the creation of a comprehensive "evaluation infrastructure", including a dedicated evaluation team of ten people serving various roles, development of extensive metrics for measuring business case parameters, technical infrastructure

> **Brief Summary: Research on Implementation of Electronic Meeting Systems**
>
> After a dominance of laboratory experiments, there is now a growing number of field studies on EMS use in organisations. Most of this research focuses on effects of this technology on meeting processes and outcome, and there are yet few studies of the process of organisational implementation of this technology. This is also due to the fact that there are relatively few organisational installations of this technology, as many companies instead use this through universities or other third-party sites. The increasing number of web-based EMS products may result in more widespread deployment.
>
> The field studies show that EMS use may result in increased productivity, broader participation and buy-in, and improved decision quality. Critical success factors in the implementation and use of this technology include organisational commitment, training and facilitation support, and dedicated meeting facilities. Conducting a cost–benefit analysis for this technology can be challenging, but is important for meeting managerial expectations and as the basis for further deployment of the technology.

development and facilitator training. The results from the evaluation period documented dramatic improvements in efficiency and effectiveness, and illustrate the value of applying a business case approach to technology evaluation for documenting the benefits from EMS technology.

Chapter 5 presents in-depth experiences from the process of implementing GroupSystems in a Norwegian oil company, as part of a portfolio of collaboration technologies. In contrast to the other case studies presented in this section, the implementation of GroupSystems in this organisation was slow, at least compared to other collaboration technologies. Problems identified in this process were recruitment of facilitators, internal marketing, and problems in quantifying intangible benefits from using the technology (Munkvold and Anson, 2001).

3.4.4 Implementation of Coordination Technologies

In Chapter 2, two main categories of coordination technologies were presented: workflow management systems and online calendar systems. Online calendars were long referred to as an example of a "groupware failure", not being widely adopted in organisations. A possible explanation for this was problems in establishing a critical mass of users due to disparity in work and benefit from maintaining these calendars (Grudin, 1989, 1994b). During the 1990s, however, this technology has become more widespread in use. Chapter 8 presents the results of a study of various factors leading to this widespread adoption.

Workflow management systems have also had a relatively slow growth in industrial use, compared to other collaboration technologies (see Chapter 2 for a discussion of possible causes for this). Several studies report problems in the process of implementing workflow systems, and in the academic literature there are still only a few examples of success stories on the use of this technology. Sceptics have pointed to inflexibility in the automated routines as well as possible misuse of managerial surveillance as potential negative aspects of this technology. Albeit somewhat slower than expected, however, the number of company installations of such systems is rising steadily, and examples of successful use are also starting to appear in the research literature. The rest of this section presents a selection of studies on workflow implementation that represent examples of both successful and less successful implementation and use of this technology.

Problems in Workflow Implementation

Workflow implementation in German industry

A group of German researchers studied six companies in different industries (transportation, telecommunication, insurance and production) which were either planning, developing or introducing workflow applications (Weske et al., 1999). All companies experienced major problems in these projects, resulting in extensive delays and in several cases also complete abandonment of the technology. Six major problems were encountered in the process of workflow application development in these projects:

- Lack of integration of organisational and technical aspects of the workflow model, as these often were worked out independently.
- Selection of workflow system at a very early stage in the project, later experiencing problems with supporting specific requirements needed in the project.
- Lack of prototyping in the development process.
- Problems in transferring the business process model into a workflow model, due to different focus and possible limitations in the workflow system.
- Resource demanding integration with legacy systems.
- Severe performance problems identified during field tests in all six companies.

To avoid some of these problems, the researchers present a reference model for workflow application development comprising six stages:

survey (identifying business process to be supported and developing a process model of this comprising both technical and organisational information), design (defining the new business process model), system selection, implementation, test phase and operational phase. Contrary to the practice observed in the industry cases, system selection according to this model is conducted after the new business process model has been designed.

Implementing workflow management on the print industry shopfloor

Adding to the problems in the early stages as described above, a much cited study also reports problems in the operational stages of workflow implementation (Bowers et al., 1995). As part of the requirements of a major tender won by a UK printing company, they had to install a workflow system for monitoring work at the shopfloor. The researchers observed how the shopfloor workers prior to the implementation applied different ways for ensuring a "smooth flow of work", such as prioritising work on the spot, reshuffling tasks among the different workers to balance the load, preparing for anticipated tasks and being able to monitor the production process through audio information in the form of alerts, regular versus irregular machine noise, etc.

When the new system was implemented, this imposed a new work model that was basically different from the one being practised among the shopfloor workers, thus disrupting the smooth flow of work and creating major overheads and obstacles in the shopfloor operations. For example, the new system identified all tasks by its job number, making it difficult to register activities related to preparation of incoming tasks. Further, the system was built around a one-to-one relationship between operators and processes, making it difficult to register the common practice of sharing processes among operators during execution. As a result, the workers had to develop different ways of working around the system and its constraints. The most extreme of these workarounds was that of one department who only entered data into the system at the end of the day based on manual notes kept as before.

Examples of Successful Implementation of Workflow Systems

Implementing workflow functionality for supporting configuration management

A field study in two companies in the IT industry describes how workflow management features are successfully used to support configuration management in systems development (Grinter, 2000). Contrary to the findings from the printing company discussed above, the formalisation and automation of work processes offered by the workflow technology here led to successful improvements of the work process compared to the previous manual methods for configuration management. This was obtained through providing support for identification and control of changes to components ("checking out/in" of software for revisions), problem management through logging of problems and process management by assigning roles to the different tasks in the development life cycles. These features resulted in the following areas of improvement:

- faster and more reliable assembly of software components to build new testable versions of the software, and improved support for locating the source of errors;
- increased awareness of the development state through logging information about changes made;
- support for prioritisation and assignment of outstanding work to developers, through integrated problem reporting facility.

In contrast to the printing case, the work model embedded in the configuration management tool corresponded well with that of the software development practices previously enacted by the developers, and the developers thus understood and accepted this model. Further, the processes automated were "right" in the sense that both users and managers supported these. The system was also flexible in that it did not automatically assign problems to individual developers. Instead, a group of people would meet and assign problems based on the problem log, and assign priorities to these. The system would then notify each developer responsible for resolving the problem, but would not require that the problems were addressed in the prioritised sequence. Finally, the supportive culture of these companies was also important for the successful use of the technology. As examples of this, the system were able to structure their workdays and schedules in an autonomous way, and managers were willing to take the developers' opinions.

Experiences from workflow implementation in an Australian IT company

The use of two Lotus Notes-based workflow systems for tender assessment and service request management in a medium-sized IT company in Australia, provided the following benefits (Atkinson and Lam, 1999):

- improved status tracking and liability;
- consistency and conformity, standardising work practices;
- improved productivity;
- improved management support through providing status information and enabling load balancing of task allocations.

However, some negative impacts related to the social interaction within the organisation were also identified. Some employees pointed out that there tended to be an over-reliance on use of the system for communication, leading to a reduction in human interaction. Further, the possibility for surveillance by management through the system was perceived to increase work pressure among employees, and they had a constant feeling of "Big Brother" watching over their shoulder.

BPR at Federal Express

The ORION project at Federal Express is another example of successful introduction of a workflow management system (Candler et al., 1996). Framed as a BPR project, this involved the implementation of an integrated system for imaging, document management and workflow, leading to major savings and efficiency gains in the company's document handling. The following factors were reported as being critical for success in this project:

- project sponsors; a senior executive sponsor identifying opportunities for innovative uses of the technology, and an operating sponsor addressing short term-issues;
- IS support; the IS department as enabler and facilitator;
- seamless integration between new and existing technology;
- staged development, utilising new technologies as they emerge;
- technology review and readiness; the implementation team proactively seeking new and relevant technologies;
- network infrastructure, offering high bandwidth for increased capacity and quality.

Brief Summary: Research on Implementation of Coordination Technologies

Despite increasing use in industry, studies of organisational implementation of workflow management systems are relatively limited in number. Most of these report on implementation failures, explained by inflexible process design, lack of correspondence with existing practices, and immature technologies. Yet, there are examples of studies that mirror the many successful applications of this technology in industry, focusing on how users have experienced major benefits in the form of improved productivity, consistency and status tracking of the work processes.

Some key issues related to successful implementation of workflow systems are in-depth mapping of the "actual" work practices to be able to develop solutions that support these, and providing the users with some autonomy and flexibility in task assignment and prioritisation.

The use of a staged approach is emphasised as a way of alleviating the difficulties of managing the massive change introduced in the organisation through this project. Further, this also allowed for periodic evaluation of risks and implementation of corrective actions.

3.4.5 Implementation of Integrated Products

This category comprises products combining features from several collaboration technologies. Three major types of integrated products were presented in Chapter 1: *collaboration product suites* such as Lotus Notes, combining various asynchronous technologies (e-mail, document management, threaded discussions, calendar and scheduling, etc.), *integrated team support technologies* including team/project rooms and desktop conferencing systems, and *e-Learning technologies*, combining various asynchronous and/or synchronous technologies.

The research on implementation of collaboration product suites is clearly dominated by studies of Lotus Notes. There is now a considerable number of such studies, some examples of which will be presented in this section. As for desktop conferencing systems there are still few studies focusing on the implementation of combined audio, video and data conferencing. As discussed related to the implementation of data conferencing, several companies seem to do without the video channel when running distributed meetings from their offices. However, the use of these technologies can be expected to grow in the future.

e-Learning technologies have only just started to become widespread, so apart from consultants' reports and vendor white papers there is little field-based research available to inform us on guidelines for successful implementation and use of these technologies. Chapter 10 presents the experiences from developing and implementing a customised MBA e-Learning program for PricewaterhouseCoopers North American Consulting Group, provided by the University of Georgia. This study is thus one of the first describing this type of implementation.

Studies of the Implementation of Lotus Notes

Being the first collaboration technology to gain widespread adoption in industry and considered the "groupware standard" for many years, the number of field studies on Lotus Notes is greater than for any other collaboration technology. The flexibility and related complexity of Notes combined with its marketing as a tool that will "transform" the organisation through increased communication and information sharing, have also attracted interest from the research community. (For a selection of "mini-cases" on organisational applications of Lotus Notes, see Lloyd and Whitehead, 1996.)

Many of the field studies of Lotus Notes implementation report problems in deployment and use of this technology to achieve the expected benefits such as increased collaboration. In the following, a selection of studies on Lotus Notes implementation is presented, that together illustrate the complexity involved in implementing a collaboration product suite such as Lotus Notes.

Cognitive and structural barriers to Lotus Notes implementation

In a frequently cited case study, Orlikowski (1992) discusses the issues related to the process of implementing Lotus Notes in a large consulting company. Over a period of 5 months she conducted more than 90 interviews with different categories of participants (consultants, administrators, technical support personnel). The technology was mainly intended for supporting information exchange among the consultants. During the period of this study, the technology was implemented in the entire organisation. However, Lotus Notes was largely used as an individual productivity tool, with extensive use of the e-mail functionality. The use of Notes to share expertise was found to be advancing more slowly.

The results from the case study suggest two categories of factors having influenced the implementation of Lotus Notes: *cognitive* and *structural* elements. The cognitive factors identified in this study include:

- limited information to the users prior to the implementation, resulting in confusion and uncertainty about the nature of the technology;
- lack of a formal implementation strategy;
- top-down approach;
- training being limited to Lotus Notes functionality, without focusing on how to use the technology for supporting collaboration;
- a resulting focus on the automation of existing routines instead of using the technology for increasing collaboration among the consultants.

Structural elements identified were related to the reward systems and policies of the organisation. The culture in this organisation can be characterised as competitive, and reward systems and promotions were based on individual achievements. This was found to be an important barrier to the use of a collaborative technology like Lotus Notes, as the incentives for sharing information were not present. Further, the hectic work pace with strong focus on revenue per employee resulted in the employees not finding time to experiment with the technology.

This study shows how the users' mental models and the organisation's structure and culture significantly influenced how Lotus Notes was implemented and used. Implications from this are that training should emphasise the collaborative nature of the technology, and that sufficient

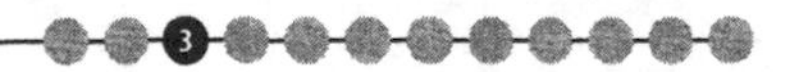

resources should be allocated for allowing the users to experiment with the technology. A limitation of this study is that it only covers the first months of use after the implementation. As noted by the author, it is possible that the use of the technology would change as the experiences with the technology increased.

Lack of effects on collaboration

A ten-month study of the effects of implementing Lotus Notes in a large US insurance company supported Orlikowski's (1992) findings (Vandenbosh and Ginzberg, 1997). Despite showing an increase in perceived usefulness of the technology as well as in perceived organisational efficiency and effectiveness, no actual increase in collaboration at departmental, divisional or corporate level in the organisation was identified. Those who were most engaged in collaboration prior to the Notes implementation continued to be so, while the technology was not found to induce collaboration among those previously not being active in this. The researchers explained this by pointing to the following negative conditions:

- no clear need for collaboration defined prior to the implementation;
- lack of understanding of the technology among the users;
- lack of sufficient training in the underlying collaborative framework of the technology;
- lack of an organisational culture supporting collaboration.

Downing and Clark (1999) argue that when assessing the impact of collaboration technologies on the level of collaboration etc., it is also important to address original expectations and anticipated outcome from the implementation. They conducted a survey among 22 US-based consultancies on their expectations and realised benefits from implementing collaboration technologies. Ninety-two per cent of the companies were using Lotus Notes, while the rest were using Microsoft Exchange. The applications ranged from e-mail and messaging, conferencing (discussion databases, electronic conferences, bulletin boards), group calendaring and scheduling, workgroup collaboration and consulting engagements.

While all 22 companies had high expectations concerning their collaboration technology implementations, they reported various degrees of success related to increased client/customer service, increased communication within the company, competitive advantage, cost savings, increase in productivity and leveraged expertise. Several companies experienced little or no benefit related to these factors. Based on their analysis of factors that may have accounted for the differences in expected versus experienced benefits, Downing and Clark (1999) present the following guidelines for implementation:

- a collaborative culture needs to be present – if not, start with smaller, less extensive pilot projects to see if such a culture can be created;
- ensure the existence of a highly visible and highly placed champion;
- define and understand baseline benefits expected – create realistic expectations and avoid broad and difficult-to-measure expectations;
- provide training for business use, focusing beyond technical know-how;
- aim for the right balance between motivating usage through management direction and encouraging experimentation and improvisation;
- increasing measures such as competitive advantage takes time – there is a need to allow time for critical mass technologies to gather momentum.

Individual interpretations of Lotus Notes

Several studies have also addressed how the integrated functionality and resulting flexibility in Lotus Notes may result in users developing different interpretations of the technology. This is found to influence both the implementation and use of this technology.

Korpela (1994) interviewed three Lotus Notes user groups in different organisations, focusing on the variation in individuals' interpretations of this technology, and the development of shared meanings. The degree of shared understanding of the technology among the users was dependent on several factors, such as information about the technology, training, "task fit", previous experiences with similar applications and time available for experimenting with the technology. The study concludes by arguing that the applications are of key importance in forming the understanding of Lotus Notes for individuals and groups. If these applications support the tasks at hand, a shared understanding of the cooperative nature of the technology is believed to emerge "in action and interaction within a time span of a few months". Implications from this study regarding implementation of the technology is that information and training should be given first, and that the users should then be allowed the possibility for experimenting with the technology in groups.

Implementation and use of Lotus Notes in Norwegian organisations

Bratteteig (1998) presents a multiple case study including analysis of the implementation and use of Lotus Notes in nine Norwegian organisations. The organisations studied were classified as project-based organisations (research institutes, trade union), stable organisations with project-oriented work styles (publisher, newspaper, police department) and large, decentralised organisations (public office, engineering organisation, computer organisation). The study is based on reports from groups of students investigating Lotus Notes usage in these organisations, and shows how use of Lotus Notes varied among the different organisations.

In all nine organisations the technology was used for supporting some aspects of collaborative work, but only two of the organisations were found to be using large parts of Notes' potential. In most of the organisations, the introduction of Lotus Notes had been stepwise, initiated by an enthusiast (e.g. a middle manager). The training in these cases had been scarce, often limited to a voluntary half-day course at the vendor. In the organisation found to be most successful in utilising Lotus Notes, however, the implementation had used a more systematic approach, involving use of a pilot group, thorough training and mandatory use.

The study also illustrates that adaptation and tailoring of Lotus Notes to local needs should not be considered an easy task, due to problems of technical integration of platforms and applications. The study concludes by pointing out that Notes often seem to be used for supporting existing practices in the organisations. Collaborative work practices should therefore be established before the technology is introduced. Further, training is essential for successful introduction and use of the technology.

A longitudinal study of Lotus Notes implementation in a consulting company

Karsten and Jones (1998) report from a three-year study of the organisational change process related to introduction and use of Lotus Notes in a Finnish computer consultancy. Over this period, the overall volume and scope of Notes usage raised steadily, and the extent of collaboration among the consultants in managing their company also increased. However, rather than ascribing this increased level of collaboration to the introduction of collaboration technology in the company, the authors present a more nuanced analysis of how this gradual organisational change was a combined result of shifting economic conditions, changes in management style, and changes in the consultants' role and work practices.

During the three-year study period, the company had three different managing directors, each representing different management styles regarding centralisation versus decentralisation and teamwork, formal management structures versus democratic and informal structure, etc. Notes was found to be adaptable to each of these approaches, both in terms of applications developed and the ways in which these were used.

In general, the authors conclude that the uptake and use of Notes in this case was more strongly influenced by aspects of the organisational context and internal social structure, than by any "intrinsic logic" of the technology.

A classification of Lotus Notes usage

Based on a review of 18 case studies of Lotus Notes implementation, Karsten (1999) conducted an extensive analysis of the relationship between collaboration and collaboration technology. She identified three categories of Notes use:

- *Exploratory, conservative or cautious use.* This involved six cases where the applications were only used to automate the existing routines, or they were very limited in scale, or no major applications had been built at the time of the study.
- *Planned and expanding use.* This was the largest group of organisations, comprising eight companies that had implemented initially restricted applications, but which had plans for expanding these.
- *Extensive and engaged use of Notes.* The third category included cases where the use of Notes was extensive and where the users took an active role in integrating Notes applications into their work. Further, the nature and amount of collaboration was changed in the cases. Only 4 of the 18 cases fell within this category.

Based on this analysis, Karsten questions the "deterministic" preconditions for increased collaboration through implementation of collaboration technology suggested in previous studies and in the trade press. Rather, she argues that whether collaboration technologies such as Lotus Notes can contribute to an increasing level of collaboration is highly contextual, depending on conscious and continued efforts to change the work arrangements related to the technology.

Alignment of collaboration technology adoption and organisational change

A comparison of the organisational adoption of collaboration technologies and related organisational change in five companies, supports the contextual view presented by Karsten (Munkvold, 2000). This study shows how successful implementation of collaboration technologies may follow different "patterns". Although adoption of the technology was seemingly most successful in cases where collaborative work practices were operational prior to the implementation, there were also examples of how adoption succeeded without this condition being present. This can be ascribed to the learning and maturation process taking place as part of the implementation process.

The relative success of the adoption in these cases shows that collaboration technology can also be used effectively in contexts where a collaborative culture does not exist prior to the implementation. Indeed, in several of the cases the move towards new collaborative work practices could not have been realized without the new technology. This is particularly relevant for distributed settings, where increased collaboration is

dependent on a technological infrastructure for communication and information exchange.

Collaboration technologies as "fragile" and "drifting" technologies

Ciborra (1996) presents a compilation of case studies on the implementation and use of collaboration technologies (mainly Lotus Notes applications) in large, complex organisations, grouped into three categories: the software industry, R&D and marketing, and the service sector. For all these organisations the implementation was part of a process of organisational transition toward new ways of working. The focus of the studies was the extent to which advanced forms of work organisations, such as teamwork, were enabled or supported by the new technological infrastructure.

Ciborra discusses the "fragile" nature of the collaboration technologies that became evident in several of the case studies. Unless the new technology is the only alternative for conducting a task, the availability of substitute media (e.g. telephone, fax, existing e-mail systems) may constitute a threat to the implementation, as users will easily return to these media in case of technical problems with the new media. The case studies also illustrate how the collaborative applications may "degenerate into traditional IS", meaning that the potential for using the technology for collaboration is not realised.

Finally, Ciborra describes the development and use of this technology as "*variable, context-specific and drifting*". He bases this on the observation that in all the cases studied, the role and function of the technology shifted by various degrees, compared to the planned objectives. This is ascribed both to the flexibility ("openness") of this type of technology, and to the improvisational nature of the type of work that is supported by this technology. The studies also showed how use of the new technologies was often restricted by the "rules" of the old organisation structure such as limited information sharing, lack of incentives for effective use of the new technology, and reservation among the users against the transparency of the work processes induced by the new technology.

An improvisational model of change management

An example of "technological drift" is reported in a study by Orlikowski (1996a, b), where the organisational change process related to the implementation of a Lotus Notes customer support system took the form of a series of incremental changes to the situated work practices, only part of which were anticipated. Based on this, Orlikowski and Hofman (1997) present an "improvisational model of change management", comprising combinations of *anticipated*, *emergent* and *opportunity-based* change. They argue that this model best represents the process of organisational

Brief Summary: Research on Implementation of Integrated Products

The review on integrated products has focused on studies on the implementation of Lotus Notes. As one of the earliest collaboration technologies, there is today a significant body of research on the organisational implementation and use of this technology. These studies show that many organisations do not succeed in realising their expected benefits of increased collaboration. This is often ascribed to insufficient user training in the collaborative potential of the technology, lack of explicit routines for effective use, and lack of incentives for information sharing. The flexible and "malleable" nature of this technology means that the individual user's interpretation of the technology becomes important for framing the use.

Further, these implementation studies also illustrate the complex change processes often associated with the implementation of Lotus Notes, spanning several years and involving "drift" from the planned objectives and a resulting need for improvisation.

transformation related to the implementation and use of flexible and "open-ended" technologies such as Lotus Notes. The key concern of the organisation in this process thus becomes to "nurture" this improvisation, by creating organisational arenas and roles that allow for continuous reflection on the change process and identification of new opportunities emerging in the process.

3.5 Overall Summary and Conclusions

The review presented in this chapter has identified a wide range of experiences, barriers and challenges related to the implementation of collaboration technology in industry. Still, the review is not exhaustive and several other studies could have been included as well.

A general limitation of literature reviews based on academic publications is the time lag before implementation of new technologies become addressed in this literature, due to the review process, etc. This means that field studies of the implementation of technologies such as instant messaging, mobile communication and web-based applications have not been available for inclusion in this review. The reader is encouraged to be on the lookout for these as well, to complement this review. Yet, many of the findings presented in this review would be expected to relate to these emerging technologies as well.

This review also illustrates the broad spectrum of technologies and applications studied in this research area, and the resulting problems in trying to develop a generic summary of findings. To help somewhat in reducing this complexity, this review has been organised according to the different categories of technologies defined in Chapter 2. In Chapter 4, a simple taxonomy of implementation factors is introduced for further classifying the findings from this review.

4

A Taxonomy of Implementation Factors for Collaboration Technologies

Bjørn Erik Munkvold

4.1 Introduction

The review in Chapter 3 presented a selection of studies on the implementation of different collaboration technologies in various organisational settings. The degree of success in these cases varied, and several of the studies reported problems related to deployment, adoption and use of the technologies. A wide range of factors influencing these implementation projects were identified. This chapter presents a simple taxonomy for categorising these factors, thus preparing the ground for the field studies presented in Part 2 of the book. The reader may then apply this taxonomy when interpreting the experiences and lessons learned presented in these studies. The same taxonomy is also used as the basis for the cross-case comparison and integration of lessons learned in Part 3 of the book (Chapter 11).

The next section introduces the taxonomy and its categories, and the rest of the chapter presents implementation factors for each of these categories, including summary tables. "Factor" should not be interpreted here in a deterministic sense, as implying clear cause-and-effect relationships. Rather, these factors have been found to potentially impact the process of implementing collaboration technologies. The nature and scope of this impact will depend on the specific implementation context, formed by the organisational context, implementation project and process, and technology characteristics. The summary tables thus only indicate possible impact of these factors on the implementation.

The review in Chapter 3 is not exhaustive, and the different factors covered here should mainly be viewed as examples of the different categories.

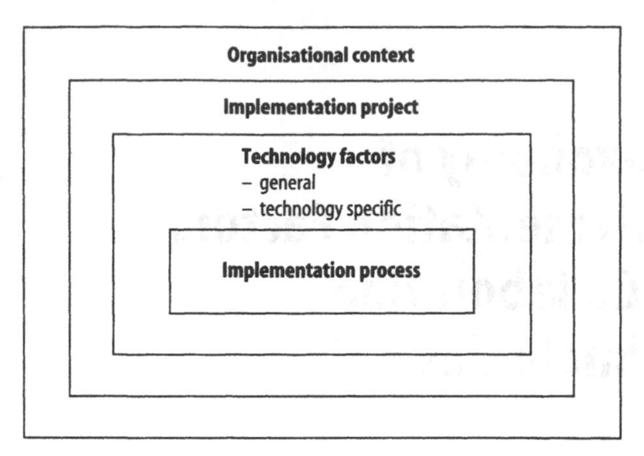

Figure 4.1 Categories of factors influencing implementation of collaboration technologies.

However, the summary tables together cover the factors most frequently reported in the literature on implementation of collaboration technology. The experiences and lessons learned presented in Part 3 of the book will complement and extend the factors presented in this chapter.

4.2 Introducing the Taxonomy

The importance of focusing on context and process when studying organisational implementation of technology is widely acknowledged (Pettigrew, 1990; Walsham, 1993). In addition, many of the factors identified in this review are related to the characteristics of the implementation project and the technology. Figure 4.1 shows the categories applied in the taxonomy for classifying implementation factors.

Organisational context factors characterise the context in which the implementation takes place. This includes both factors related to the organisation's external environment, such as characteristics of the industry and relations to third parties (vendors, partners, customers, etc.), and internal characteristics of the organisation such as culture, previous experience with collaboration and IT competence. *Implementation project* factors are related to the organisation and conduct of the implementation project, for example user training and establishing a support infrastructure. The *technology* category is divided into factors that are more or less general for collaboration technologies, and factors that to some extent are identified as specific to certain technologies. The latter will be classified in sub-categories according to the framework presented in Chapter 2. Factors in the *implementation process* category characterise the nature of the implementation process, such as time frame of the implementation, nature of the change process, top-down or bottom-up approach, etc.

As illustrated in Figure 4.1, these different categories are clearly interrelated. The organisational context will frame the background and purpose of the implementation project, which in turn will frame the choice of technology and its implementation. Together, these categories frame the nature of the implementation process. Below, each of the categories is discussed in more detail.

4.3 Organisational Context

The findings from previous field studies show how the organisational context as defined in this taxonomy can influence the organisational implementation of collaboration technology in several ways.

An organisation's "readiness" for adoption and use of collaboration technology is found to be dependent on factors at several different levels, such as the existing degree of collaborative work practices in the organisation, the users' felt need for technology support in their work, the existing IT infrastructure and IT competence and experience in the organisation, and support from top management. Several studies also argue that the existence of a "collaborative culture" is a precondition for successful implementation and use of collaboration technology. The extent to which the organisation's reward system and policies stimulate collaborative versus individualistic work is often taken as an indication of the collaborative culture in the company (Orlikowski, 1992).

However, other authors argue that this last point regarding collaborative culture reflects a deterministic stance, implying that successful adoption and use will occur "only" if a collaborative culture exists (Karsten, 1999). The study by Karsten and Jones (1998) on Lotus Notes implementation in the Finnish consulting company illustrates how other aspects of the organisational context such as recession in national economy, management style, and changes in roles and work practices may exert stronger influence on the implementation of collaboration technology than the existence of a collaborative culture. The case studies of the alignment of collaboration technology adoption and related organisational change presented in Munkvold (2000) also show how successful adoption of collaboration technology can take place in organisations without previous collaborative work practices, through a process of learning and maturation. In these cases, the technology may actually stimulate new collaborative work practices, for example those related to distributed teamwork.

Few of the studies in the literature review in Chapter 3 explicitly address the influence of external relations on implementation. The field studies in Part 2 of the book will illustrate how relations to customers, industry partners, vendors and other third parties may exert vital influence on the implementation project.

Table 4.1 lists organisational context factors, and their possible effects on implementation.

Table 4.1 Implementation factors related to organisational context

Factors	Possible effects on implementation
• Existing degree of collaborative work practices in the organisation	Existing collaborative work practices may have a positive effect on the users' receptivity for collaboration technology.
• Users' felt need for technology support	A felt need among the users has a clear positive effect on adoption of the technology.
• Individualistic versus collaborative culture	Organisations with a highly individualistic and competitive culture may face greater challenges in adoption of collaboration technologies than organisations already focusing on collaboration.
• Reward systems and policy	These structural elements are important means for stimulating collaboration and related use of collaboration technology in the company.
• Top management support	Top management support is important for providing organisational "legitimacy" to the implementation and for gaining access to adequate resources.
• Management style	Management style can impact the implementation and use of collaboration technology. However, collaboration technology can be adapted to serve different styles, and does not automatically support more collaborative and decentralised/democratic approaches.
• Existing IT infrastructure	Collaboration technologies require a basic IT infrastructure. The implementation project needs to take into account any necessary upgrades in this infrastructure.
• Existing IT competence	Lack of internal IT competence in the organisation may be a barrier to effective implementation. On a short range, vendors and consultants can provide this, but the organisation needs to build internal competence for future maintenance and support.
• Economic conditions	Economic conditions such as recession in national economy and fluctuations in market conjunctures may impact the implementation in different ways. For example, it may result in budget cuts for the implementation, or it may lead to increasing focus on how to make organisational practices more effective through collaboration technology.

4.4 Implementation Project

Several studies in the review report of shortcomings in the formal planning and organisation of the implementation project, stating the need for a formalised implementation strategy. Recurring elements in such a strategy are:

- Creating a cross-disciplinary implementation team.
- Communicating sufficient information to the users to establish realistic expectations. This should also include presentation of the vision/ rationale for implementing the technology.
- Using pilot groups with people who have a real need for collaborating.
- Providing adequate training in the technology, focusing on collaborative aspects and how to incorporate the technology in the daily routines.
- Establishing a supportive infrastructure, providing follow-through support and troubleshooting systems problems quickly.
- Establishing incentives for stimulating user adoption of the technologies.
- Defining routines for effective use of the technologies.

The use of a formal implementation strategy including the elements specified above still does not guarantee a successful implementation. Several studies illustrate the limitations of a planned approach for implementing collaboration technologies (Ciborra, 1996). The "drifting nature" of flexible collaboration technologies such as Lotus Notes creates a need to allow for improvisation to be able to exploit new possibilities emerging in the project, as illustrated in the model of planned, emergent and opportunity-based change (Orlikowski and Hofman, 1997).

In line with this, several point to the need for letting the users experiment with the technology, to discover new ways of incorporating this in their workday. This experimentation must be balanced against prescribed routines for best practice, supported by managerial mandate.

Finally, the importance of one or more project champions is generally acknowledged. Champions are often associated with formal power based on a management position. This may be necessary for mobilising sufficient resources and managerial mandate to be able to influence adoption decisions. However, individuals may also become champions on the grounds of their domain competence and recognition in the organisations. Charisma is also an attribute that is often associated with champions. Schön (1963) characterises a product champion in this way:

> Essentially, the champion must be a man willing to put himself on the line for an idea of doubtful success. He is willing to fail. But he is capable of using any and every means of informal sales and pressure in order to succeed ... they identify with the idea as their own, and with its promotion as a cause, to a degree that goes far beyond the requirements of their job (p. 84).

Table 4.2 Implementation factors related to the implementation project

Factors	Potential effects on implementation
• Formalised implementation strategy versus improvisation	A formalised implementation strategy has a positive effect on project management, including scheduling, resource allocation and coordination. However, experience shows that some room for improvisation is needed.
• Composition of implementation team	An implementation team with a right blend of technical competence and business understanding creates the required "socio-technical balance" needed for successful implementation.
• Information to the users	The information provided to the users has an important bearing on their perceptions (mental models) of the collaboration technology and its potential.
• Users' expectations	Realistic expectations towards the new technology are important to avoid any frustration and disappointment among the users. Potential benefits of the technology should be communicated to the users, but without "overselling" it.
• Composition of pilot groups	Pilot groups without a real need for technology support may fail to document the potential benefits. The members of the pilot groups should be selected on the basis of their need for collaborative IT support.
• User training	Lack of adequate training is a recurring factor in implementation failure. The training needs to include an explicit focus on collaborative aspects.
• Establishing a supportive infrastructure	Some form of support infrastructure is important to handle problems early and thus avoid user frustration.
• Project champion(s)	Access to one or more project champions has proven instrumental to implementation success.
• Incentives for stimulating user adoption	Establishing clear incentives may stimulate adoption of the technology. This could be in the form of improved working conditions for the individual employees, and/or bonus schemes for increased productivity.
• Predefined routines versus user experimentation	Clear guidelines and routines may increase the effect of the technology. This should be balanced against giving the users room to experiment with the technology, to come up with new and creative applications.

Several of the field studies in Part 2 of the book illustrate the important role played by such champions in the implementation.

Table 4.2 summarises the implementation project factors.

4.5 Technology-related Factors

This category includes all factors related to characteristics of the technology. These are divided into factors that can be regarded as more or less general for collaboration technologies, and factors that to some extent are specific for one type of technology. The latter are classified according to the categories introduced in Chapter 2.

4.5.1 General Factors

Grudin's (1994b) challenges, presented in Section 3.3, are examples of generic factors related to the implementation of collaboration technologies. Perceived disparity in work and benefit, disruption of social processes, problems related to exception handling and unobtrusive accessibility of the collaborative tools are all important factors that can impact the implementation process. However, the degree to which these factors may influence the implementation may vary for the different types of technologies, and will also depend on the organisational context.

The maturity or stage in development life cycle of the technology is also an important factor. Immature technologies may easily lead to problems with stability and performance, and result in distrust in the technology among the users. Another related issue is the compatibility with existing technological infrastructure. In general, collaboration technologies may be "fragile" in that the users can easily fall back to former work practices and related technology support (or lack of such) in case of problems or frustration with the new technology (Ciborra, 1996).

Another form of compatibility is the degree of correspondence with existing routines. The implementation of collaboration technologies may result in that work practices and routines that previously have been of a "tacit" nature, now are made explicit and "transparent" in the sense that activities can be monitored by management and fellow workers. This again may result in increased accountability, and a corresponding "fear of exposure" among the users.

Table 4.3 summarises the "general" technology-related factors.

Table 4.3 Implementation factors related to collaboration technology "in general"

Factors	Possible effects on implementation
• Critical mass	Establishing a critical mass of users is crucial for collaboration technologies where the users' benefits are dependent on universal adoption.
• Disparity in work and benefit	Perceived disparity in extra workload and benefit induced from the technology may represent a barrier to user adoption.
• Disruption of social processes	Technologies that represent disturbances to the often tacit social processes risk facing user resistance.
• Exception handling	Exceptions to the formal routines occur frequently in the day-to-day work practices. Some flexibility should be built into the systems, to accommodate for these exceptions.
• Unobtrusive accessibility	Some collaborative tools are not used as frequently as other office support tools. By offering seamless integration with the user's standard work tools, the collaboration tools also accommodate more infrequent use.
• IT maturity	Immature technology can create problems with stability and performance of the solution, resulting in project delays and distrust among the users.
• Compatibility with existing technologies	Technical incompatibility can result in project delays and frustrated users.
• Compatibility with existing routines	Compatibility with existing routines means less "friction" in user adoption. However, some implementations will aim at changing these routines.
• Fragile nature of collaboration technologies	In case of problems with a new collaboration technology, users may easily abandon this in favour of existing, substitute technologies more familiar to them.

4.5.2 Implementation Factors Related to Communication Technologies

In addition to generic factors such as critical mass and technological compatibility, the development of routines for effective electronic communication is important for successful use of these services. This involves routines for distribution of document attachments, "social protocols" for communication frequency and response time, and general norms

Table 4.4 Implementation factors related to communication technologies

Factors	Possible effects on implementation
• Routines for electronic communication	Such routines may contribute to effective use of the services, and reduced information overload.
• Social protocols for communication frequency and contents (netiquette)	Important for building relationships in electronic communication, and avoiding misbehaviour.
• Bandwidth and image quality	Critical factors for videoconferencing systems.

for acceptable communication. For videoconferencing technologies, bandwidth and image quality are also important factors influencing user adoption. Table 4.4 lists implementation factors related to communication technologies.

4.5.3 Implementation Factors Related to Shared Information Space Technologies

Several factors were identified related to the implementation of shared information space technologies, such as document management systems and data conferencing/application sharing. The transition from paper-based to electronic document handling may create challenges related to the development of new routines (Bowers, 1994). Increasing visibility of the document production process combined with the ability for sharing documents can result in dilemmas of responsibility and ownership of information at various stages in the process. Another issue is the lack of support for mobile users.

Important implementation factors identified related to knowledge repositories include the need to establish functional routines and search mechanisms for navigating the vast information spaces. Effective content management also requires clear roles and responsibilities for maintenance and quality assurance of the information contents.

For data conferencing, several factors are important for obtaining effective use. Adequate training and use of distributed facilitators are recommended for avoiding delays and problems when running distributed meetings. In addition, development of routines/protocols for structured use of the application sharing functionality is necessary to avoid problems related to "screen management" and turn-taking. Finally, technological limitations are still found to influence the use of this service, in the form of inadequate audio quality.

Table 4.5 lists implementation factors related to shared information space technologies.

Table 4.5 Implementation factors related to shared information space technologies

Factors	Possible effects on implementation
Document management systems/Knowledge repositories	
• Increasing visibility of document production process	Transition from paper-based to electronic document handling makes the document production process more transparent. This requires an analysis of possible changes in the temporality of work routines, such as related to publication and distribution of documents.
• Ownership and responsibility of information	Sharing electronic documents may raise new issues related to ownership and responsibility for the information in its various production stages.
• Support for mobile users	Limited access to digital documents for mobile users may represent a challenge in the implementation of these technologies.
• Effective search mechanisms	Critical for effective use of knowledge repositories. For organisation-wide databases, there may be a need for developing a thesaurus.
• Roles and responsibilities for content management	Effective content management using document management systems requires new roles and responsibilities for maintenance and quality control.
Data conferencing/Application sharing	
• Distributed facilitators	There is a growing attention to the importance of this role to ensure effective communication "flow" in distributed meetings.
• Technical support	Dedicated technical support can eliminate start-up delays and problems in distributed meetings. This function can also be fulfilled by the distributed facilitators.
• Routines/protocols for structured use of application sharing (screen management and turn-taking)	Such routines/protocols are needed to avoid "chaos"and ineffective use of application sharing.
• Audio quality	Limited audio quality may restrict the use of integrated audio and data conferencing.

4.5.4 Implementation Factors Related to Electronic Meeting Systems

Research on the implementation of EMS technologies illustrates the importance of an "electronic meeting infrastructure", comprising both dedicated electronic meeting rooms and trained meeting facilitators. The facilitator's ability to effectively match the tools in the system with the

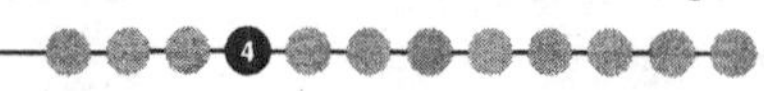

Table 4.6 Implementation factors related to electronic meeting systems

Factors	Possible effects on implementation
• Dedicated electronic meeting rooms	Co-located, electronic meetings require dedicated meeting rooms with adequate IT infrastructure. This may be a significant investment for a company.
• Access to trained facilitators	The meeting facilitator is instrumental for successful electronic meetings. He or she is responsible for planning and running the meeting, and processing the meeting report.
• Matching EMS tools with meeting tasks	Using the right EMS tools for the meeting activities is vital for the process and outcome of an electronic meeting. This is specified in the meeting agenda prepared by the facilitator.
• Balancing electronic and verbal interaction	Electronic meetings require a balance of electronic and verbal interaction to be effective. The facilitator manages this balance.

tasks at hand, and defining the right balance between electronic and verbal communication, is of critical importance for effective use of these systems. In general, the implementation of these systems requires a fine-tuned "socio-technical balance" (Bikson and Eveland, 1996). Table 4.6 lists factors related to electronic meeting systems.

4.5.5 Implementation Factors Related to Coordination Technologies

The studies reviewed in this category represent examples of both successful and less successful implementation and use of workflow management systems. The degree to which the new, automated work process enabled by this technology corresponded with the users' needs and their model of work proved instrumental for the outcome of these projects. Critiques of these systems point to the lack of flexibility and possible "intrusive" nature of these systems, providing features for management control and monitoring of work tasks ("Big brother watching"). The studies reviewed, however, also include examples where the employees perceive the workflow management system to be of great support to their work, such as in the case of configuration management in the IT industry (Grinter, 2000). This study also shows how flexibility in the use of these systems can be obtained by making the employees handle job assignment and prioritisation themselves.

Transferring the business process model into a workflow model, including integration of organisational and technical aspects, is a primary concern. A related question that needs to be resolved is the stage to select the actual workflow product to be implemented. Another important

Table 4.7 Implementation factors related to coordination technologies

Factors	Possible effects on implementation
Workflow management systems	
• Transferring business process model into workflow model	The workflow model needs to incorporate both organisational and technical aspects of the business process.
• Correspondence with users' model of work	Imposing new work models that do not correspond with the users' model of work may disrupt the "smooth flow of work" and lead to user resistance.
• Flexibility in process	Necessary for exception handling and allowing some user autonomy in job allocation and prioritisation.
• Timing of selection of workflow product	The workflow product should not be selected until after the new business process model has been designed, to assure that the product meets the requirements in full.
• Integration with legacy systems	Important but often challenging and resource demanding task in workflow implementation.
• Management surveillance	Potential risk of misuse for control purposes may result in users being sceptic. It is important to deal with this up front, to reassure users.

factor is the integration with legacy systems in the organisation. Table 4.7 summarises factors related to workflow management systems.

4.5.6 Implementation Factors Related to Integrated Products

By definition, these technologies comprise the functionality of several of the other categories. Thus, they also "inherit" implementation factors from these, as well as bring about some new ones.

The studies reviewed in this category were all on implementation of Lotus Notes. Due to the flexible functionality of integrated products such as Lotus Notes, the users' individual interpretations (or mental models) of the technology have been found to be influential for the way the technology is adopted and used. Thus, the information and training provided to the users are of vital importance for successful utilisation of integrated products. Several therefore also argue for the need to provide explicit guidelines for use. Others again point to the need for allowing time for the users to experiment with the technology, to explore new ways of utilising the technology in their daily work practices. The main challenge is to find the right balance between motivating usage through management directives and encouraging experimentation and improvisation (Downing and Clark, 1999).

Table 4.8 Implementation factors related to integrated products

Factors	Possible effects on implementation
Lotus Notes	
• Users' individual interpretations (mental models) of the technology	The users' interpretations of the technology frame the scope and effectiveness of its use. Explicit training in the collaborative features of the technology is important for demonstrating its potential to the users.
• Balance between management directives and user experimentation and improvisation	Some guidelines for "best practice" are needed to ensure effective use. This must be balanced against the need for allowing users to experiment with the technology to come up with creative applications.

Table 4.8 presents implementation factors for integrated products, exemplified by Lotus Notes.

4.6 Implementation Process

In addition to factors related to the "formal" implementation project, the review identified several factors characterising the technology adoption and diffusion process in the organisations. Establishing a critical mass of users was stressed as an important implementation factor for collaboration technologies. In addition to characteristics of the technology, this also tended to depend on social influence mechanisms such as peer pressure, and the gradual learning and experience building among the users. More often than not, the time span of the implementation process was reported to be longer than expected at the outset, due to various barriers encountered in the process. These barriers were often a result of some form of mismatch between elements of the organisational context and characteristics of the technology.

The studies reviewed also show how the implementation of the technologies may follow different "trajectories". Examples of both top-down and bottom-up approaches were found. A top-down approach is often characterised as not being sensitive to local needs and concerns and, thus, often perceived by the users as management dictate. A bottom-up approach better secures support and commitment among the users, but may lack a clear overall vision and the necessary coordination provided by managerial mandate. The implementation process, however, often emerges without any clearly defined approach at the outset. For example, in several cases of Lotus Notes implementation the first initiative originated from middle management, then spread to other parts in the organisation creating an organisational "pull" for this technology, until finally

Table 4.9 Implementation factors related to the implementation process

Factors	Possible effects on implementation
• Top-down versus bottom-up approach	A top-down implementation approach may ensure a coordinated process guided by an overall vision, but may face user resistance due to lack of adaptation to local needs and practices. A bottom-up approach may result in greater "buy in" from the users, but may lack coordination and strategic vision. When possible, a "combined approach" is recommended, stimulating bottom up adoption guided by strategic vision and central coordination.
• Social influence mechanisms	Social influence mechanisms such as peer pressure and "word of mouth" can be more influential on user adoption of a new technology than any planned approach. The implementation team should try to capitalise on this through appointing superusers and "technology ambassadors" in the organisation.
• Implementation barriers resulting from conflict between organisational context and technology characteristics	Most implementation projects encounter unforeseen barriers threatening the project. The implementation team must deal with these as early as possible, and try to eliminate any misfit between technology and organisational context.
• User learning and adaptation	The implementation cases show that users generally are able to adapt to changing work practices and use of new collaboration technology – however, this is a gradual learning process that may take long.

being lifted to the strategic level and formalised using a top-down implementation approach. Again, the challenge becomes striking the right balance between these two approaches that best fits the actual implementation context.

Table 4.9 outlines factors related to the implementation process.

4.7 Summary

In general, implementation of collaboration technologies can be seen as a variant of the broader area of IT implementation. Thus, IT implementation factors of a more general nature, such as top management support, user involvement, adequate training, etc., should be expected to be important for implementation of collaboration technology as well. In addition, the review in the previous chapter and the field studies presented in Part 2 of the book, also provide several examples of more specific factors related to each type of collaboration technology.

The taxonomy presented in this chapter is an attempt to categorise and classify the many different factors that may influence the organisational implementation of collaboration technologies. Appendix C presents the complete taxonomy, integrating the summary tables for the different categories.

Part II

Lessons from Industry

5

Implementing a Portfolio of Collaboration Technologies in Statoil

Bjørn Erik Munkvold and Bjørn Tvedte

5.1 Introduction

The focus of this chapter is the implementation of collaboration technologies in Statoil, a Norwegian oil company. Over a period of ten years, this company has gradually developed an infrastructure of collaboration technologies for supporting its work processes. This includes document management and workflow, intranet, electronic meeting support systems and videoconferencing. The case study uncovers the many challenges in the adoption of these technologies, and the continuing work of developing effective work practices related to the technology. Changing the work habits towards more collaborative behaviour has proved difficult, and IT tools supporting individual production are still more popular and widespread in use than collaboration tools such as shared databases. Further, limited coordination among the various internal groups supporting the different technologies has also acted as a barrier in this process. To improve this situation, Statoil launched a major project for integrating the further activity related to their portfolio of collaboration technologies. The main elements in this strategy are presented, together with further challenges related to the implementation of the strategy.

The chapter is based on the experiences and observations of the second author during his participation in the implementation projects described, as well as interviews with other key actors in the implementation projects and internal project documentation.

5.2 Presentation of Statoil

Statoil is a Norwegian oil company with approximately 17,000 employees and an operating revenue of over US$21.7 billion in 2000. The organisation

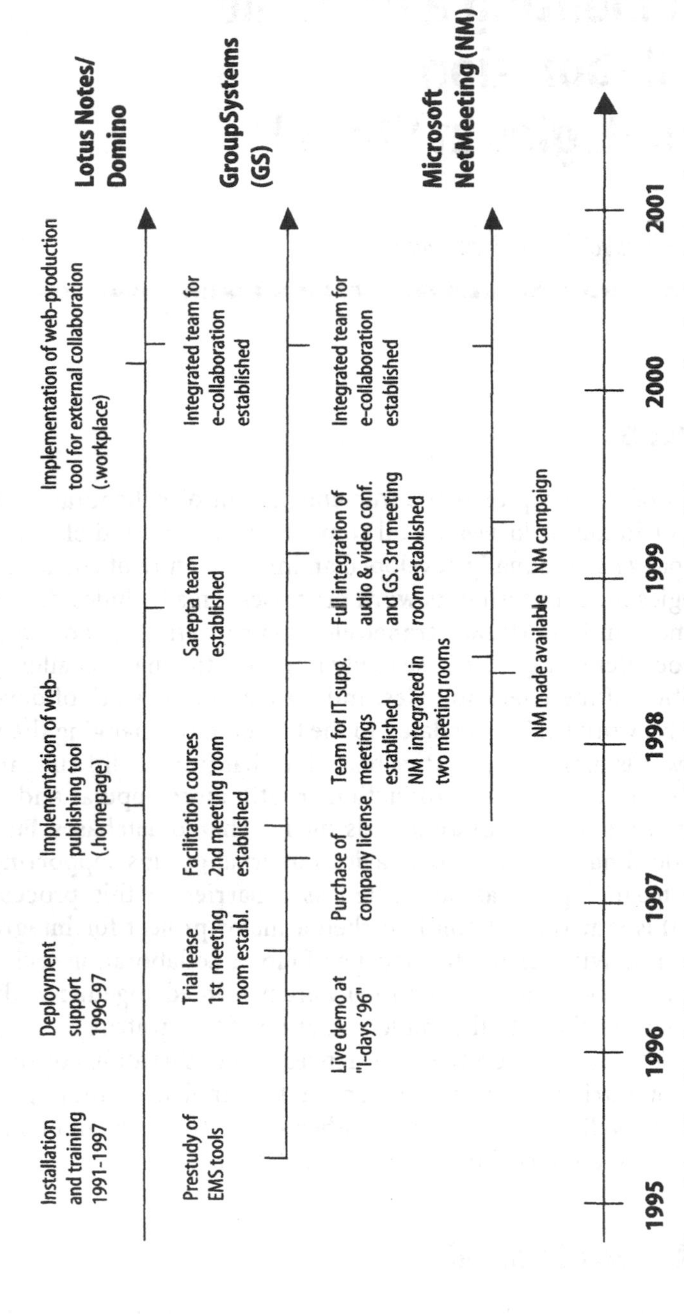

Figure 5.1 Timeline indicating major events in the implementation of collaboration technologies in Statoil.

comprises 40–50 different sites, including offshore platforms and operations in more than 20 countries. The geographic distribution makes Statoil's operations coordination-intensive, and the company is an advanced user of different IT applications for supporting communication and collaboration. With a full company licence, Statoil is one of the world's largest users of Lotus Notes/Domino, applying this for workflow and document management capabilities. Statoil also makes use of IT for supporting different forms of meetings. Through combined use of Lotus Notes/Domino, GroupSystems, Microsoft NetMeeting and audio- and videoconferencing, they have established the basis of an "anytime/anyplace" meeting infrastructure.

Statoil IT is the organisation's central IT unit, responsible for adapting and delivering IT services to internal customers in the company. The unit has about 450 employees and is represented at all major Statoil sites. About 60 of these constitute a dedicated support staff located in the different organisational units. Statoil IT also has extensive collaboration with IS/IT coordinators in each organisational unit. All activities related to the development of new IT products and services in Statoil IT, including purchase and adaptation of commercial systems and related training and support, is conducted through the form of a standardised product development process. This process involves three decision stages:

D1 approval of the establishment of a new product (initiation of project)
D2 approval of the recommended solution/technology for the product
D3 approval of the finalised product.

At each of these stages the new product is subject to approval by a Product Council comprising managers from the different areas in Statoil IT responsible for supporting the internal customers in the major sectors in Statoil.

5.3 Implementation Trajectories for Collaboration Technologies in Statoil

Over the last decade, several collaboration technologies and applications have been introduced in Statoil. Figure 5.1 presents a time line indicating major events in the implementation of three main technologies: document management and workflow (Lotus Notes/Domino), electronic meeting support (GroupSystems) and desktop conferencing (Intel ProShare, Microsoft NetMeeting).

As illustrated in the figure, the implementation of these technologies has developed gradually, from first being introduced and marketed as separate technologies to becoming more integrated. In the following, a brief description of the implementation trajectories for these three technologies is provided.

5.3.1 Implementation of Lotus Notes

Background and Scope

The first initiative for using Lotus Notes in Statoil came in 1991. Some managers who had been introduced to Lotus Notes were enthusiastic about the potential of this technology, and started internal marketing of the system. At this time there was no formalised goal for the implementation of Lotus Notes, but merely a notion that this concept was interesting. Some successful pilot projects were conducted to examine how Lotus Notes could be used, but without any overall coordination.

In 1994, Exploration and Development, one of the largest units in Statoil, conducted a cost-benefit analysis for implementation of Lotus Notes, identifying "radical potential for improvements". By the end of 1994, the unit granted money for a formalised implementation project. This started in April 1995, involving the four largest organisational units in Statoil. Statoil IT developed a toolbox of seven standard Lotus Notes applications supporting general asynchronous collaboration and coordination, shown in Table 5.1.

The implementation also included Lotus Smartsuite, Lotus' collection of office support applications (word processing, spreadsheets, etc.). Since then the portfolio of Lotus Notes applications have been continuously developed. During 1998–99, the tools for project support, document management and electronic archive were merged into a common solution called *Sarepta*, offering an "electronic project room" comprising modules for production of documents and related distribution/workflow

Table 5.1 Standard Lotus Notes tools implemented in Statoil

Tool	Application areas
Document management and workflow	Development/production, storage and sharing of information in projects, functions and other administrative processes
News and bulletin boards	Storage and sharing of general information (news and announcements) in organisational units/functions
Discussion	Development/production, storage and sharing of information in discussion groups and networks
Electronic archive	Administration, storage and sharing of archival information
Governing documents	Development, storage, sharing and continuous improvement of governing documents
Group calendar	Personal and group coordination
E-mail	Personal and group coordination

management (*Sarepta Arena*) and archiving functions (*Sarepta Archive*). As part of this, a dedicated Sarepta team was also formed with responsibility for installation of the Sarepta tools and related support.

The functionality for web integration offered by the Lotus Domino platform also formed the basis for implementation of new intra and extranet services. A standard web publication tool termed *.homepage* was implemented in 1998, enabling structured publishing of documents and news generated in Lotus Notes on the intra- and extranet. And in 2000, an application termed *.workplace* was made available as a standard tool for production and storage of documents through the web interface, especially intended for supporting collaboration with external partners. In general, over the past two years there has been an increasing focus on the development of Domino/web-based information portals.

Adoption and Diffusion

Although there was a request from top management in Statoil that the organisational units should adopt Lotus Notes, the final decision was discretionary. The management in each main business unit decided whether to adopt the system, and could also select which of the standard tools that should be implemented. The adoption can therefore be characterised as a sales process, lasting more than a year and with some units holding back more than others.

The implementation project was coordinated by a department in Exploration and Development, with Statoil IT as the operating unit. Statoil IT mobilised a large operation for implementing the technology, mainly using personnel from the first and second order line support. During peak periods, however, a larger number of personnel from all Statoil IT were mobilised for conducting user visits. According to the project leader, this was important for being able to maintain the schedule. External consultants were used to some extent in the work of making the applications ready for installation. But Statoil IT had the main responsibility and also did a great deal of this work. The consultants did not participate in any advisory functions. Statoil IT was responsible for all systems administration, including maintenance and user support.

Statoil IT put emphasis on providing information and training to the users, and the implementation project comprised the following phases:

- Preparation and planning with each organisational unit where Lotus Notes was to be used, including a discussion of which standard tools the unit wished to start with. The network of IS/IT coordinators was used as the customers' focal point of contact in each unit. First information was given to this network, then to the network representative in each unit together with management, leading to a decision of which tools in the standard package they wanted to use.

- Installation of the technology, giving users access to Lotus Notes and inviting them to introductory meetings with presentation and demonstration of Lotus Notes.
- Individual training in each user's office, lasting one hour.

Different roles were also defined for the administration of Lotus Notes in each unit:

- *Database administrator* – responsible for preparing the database for use, and seeing to that the right persons have access to the information in the database. Also responsible for the training in the local unit.
- *Editor* – responsible for the information put in the database, and updating and removing old information.
- *IS/IT coordinator* – assesses new standard tools developed, orders the tools and prepares for use in the unit, and maintains contact with the vendor/unit responsible for the product.

There was some initial resistance to adoption in the first stages. The Lotus Notes tools were replacing existing systems, such as the MEMO mail system and an archiving system, and this created concern among those responsible for these technologies. It was therefore important for the implementation team to collaborate with these persons. The users were also comparing the functionality of the new technology with that of the existing, and differences here could inhibit adoption. As expressed by the project leader:

> The main problem was the e-mail system in Notes. It didn't have all the functions that the old system had (in the beginning, when we were marketing it). The users missed some of these functions, and were focusing much on how they did not have the same functions as in the old tools. Those functions came later, during the same year.

There were also some technical problems related to the integration of new and existing technologies. For example, during a first transition period it was necessary to provide two-way integration between the old and the new e-mail system. This interface was unstable, creating resistance from the users and many complaints that this was not good enough. After one year, the old system was phased out completely.

Despite these problems, the installation and training was accomplished according to the initial schedule. The number of Lotus Notes users grew rapidly, from approximately 1,000 in 1994 to more than 18,000 in 1997 (Monteiro and Hepsø, 2000), implying full coverage among Statoil's employees. Rather than being a result of a top management initiative, the project had more the character of a bottom-up approach. The project leader puts it this way:

> Notes came as a good idea, spread among some enthusiasts who managed to market the product internally. Gradually it became a demand that everyone wanted Notes. The product in many ways sold itself.

The network of IS/IT coordinators also had an important role in facilitating user acceptance of Notes, acting as local champions in the acceptance stage. The project leader expresses it this way:

> This has been a success criterion for us: To get the IS/IT coordinators motivated, and become local champions in the implementation. This was really decisive for the progress we had. The IS/IT coordinators who were more sceptical were holding back, but we did not feel that there were any special problems. Most of them were willing to collaborate – most become willing to collaborate when they get information. If you share information with someone it is easier to establish trust.

Varying Acceptance of Different Tools

Despite the rapid diffusion of Notes among the employees, the actual adoption and use of some of the tools was slower. While applications that can be characterised as support tools (e-mail, group calendar, meeting room reservation, archiving, etc.) were used extensively, applications requiring input from each user in a common database were slower in adoption. In general, tools for distribution of information, enabling "one-to-many" communication, have been most widely accepted. The document management and workflow tool requiring each user to produce input to the shared databases was used less, although this application was believed to have the greatest potential for improving efficiency. As stated by the project leader:

> All the standard tools can be characterised as support tools, apart from the workflow tool. These are tools that support basic processes, e.g. archiving and coordination. They are not production tools. If they are used right, they may contribute to collaboration and information sharing, but they only influence each individual to a small degree. You are not dependent on them for doing the things you need to get done. Then you instead use Notes mail and the office support tools locally on your PC. You do not have to produce in the databases. When it comes to the standard tools, any major changes in the way people work is not expected. It is just a system that makes it easier to rationalise some processes.

As a result, Lotus Notes was still not exploited to its full extent. The implementation project had succeeded in delivering a set of support tools that were being used extensively for coordination and distribution of information across the whole organisation. However, the internal customers responsible for the implementation in the different units felt that the potential of utilising the system for making information sharing and collaboration more efficient was not fully realised.

Utilisation Support

To advance further in this process, a new project was initiated, entitled *Utilisation support*. The aim of this project was to motivate and discipline the users to share information and use the common databases. The project was also informally named as "the paperclip hunt", referring to the paperclip icon used in Lotus Notes as the symbol for e-mail attachments. An important part of the project was to change the users' habits of distributing documents as e-mail attachments and instead use the common databases developed for this purpose, to enable general access to these information resources. Central in this project was also the use of a market metaphor for distribution of information, forcing information suppliers to focus on the quality of the information, and making "information customers" responsible for seeking information. Statoil IT had the role as the market entrepreneur, offering the necessary tools. As part of this project, guidance for adapting the use of the system to local conditions in each unit was also developed. Further, the project also focused on coordination and standardisation of the large number of databases in Lotus Notes that had been developed after implementation of the system.

This work proved to be difficult and challenging, and many of the same problems related to information sharing and effective use of the tools can still be seen to exist today. The implementation of Lotus Notes provided useful experience and insight into the possibilities and limitations of this product for supporting collaboration. The overall focus in this implementation project was on providing support for collaboration processes, rather than "pushing" a specific technology. Thus, Statoil IT found that there were several central processes and aspects of collaboration that were not supported by the standard tools in Lotus Notes. Obviously, one of these were meetings. As a consequence, they realised that to get further in this they needed to expand their technology portfolio to also include electronic meeting support tools.

5.3.2 Implementation of Electronic Meeting Systems (EMS)

Adoption Process

In 1995, the IT department approved a pre-study of potential meeting support technologies, initiated by the second author (at the time working in the Lotus Notes implementation project). In January 1996, GroupVision, a consulting company marketing GroupSystems, introduced Statoil IT representatives to another Norwegian customer already using the technology. A demo for the managers in Statoil IT was held at the location of this company, creating an "overwhelmingly positive attitude" towards the tool. This was followed up with a live demo at the annual IT seminar in

Statoil, the "I-days '96", with a large stand and a demo room set up with 15 participant workstations and a large interactive screen (LiveBoard). This demo was also described as a success, exposing the product to a lot of people and also enrolling new allies such as the person responsible for the unit for new product development in Statoil IT. Despite this, the initial proposal to purchase GroupSystems was rejected by the Product Council in the IT department, allegedly due to the costs. However, the proposal was further developed and later that year gained approval for entering a half-year lease arrangement with GroupVision, as a test period.

Soon after this the first meeting room was established at the Statoil headquarters in Stavanger, including 14 participant workstations running GroupSystems on a separate LAN and a large screen. The workstations were former generation PCs with a docking station and large keyboard, thus representing a cheap but functional solution. The second author started working full time as a GroupSystems facilitator, and also took on the responsibility for conducting internal marketing of the services and running demonstrations. However, the marketing was only local to this specific site, as this created more than enough demand for one facilitator working full time. An evaluation scheme in the form of a questionnaire to be completed by the participants after each meeting was designed. The results from this evaluation were positive and were used as the basis for a proposal to purchase the technology at the end of the test period. This was approved, and in March 1997 Statoil purchased a full company licence of GroupSystems. During summer 1997 several facilitation courses were run, with a consultant from GroupVision as instructor. At the same time, the second meeting room in Trondheim was established, and a second facilitator was employed so that there was one at each meeting room. In addition, mobile GroupSystems facilities including a facilitator were hired on several occasions for use in larger meetings both internally and externally. In 1998, mobile use of GroupSystems was included in the product portfolio offered by Statoil IT, although they do not have this equipment in-house.

Despite this progress in establishing an EMS infrastructure, the activity dropped somewhat in the following period. The process of obtaining the approval had been lengthy, with continuous challenges from internal competition and political struggle. For example, there was a group in Statoil's research unit working with collaboration technology, but with a more exploratory approach to the different technology areas. This group was also selling their services to internal customers in Statoil, thus acting as a "semi-competitor" to Statoil IT in certain areas. Being good at networking and relation building, this group had also gained an influential position in the company towards other potential adopters in management. This group was more reluctant towards GroupSystems, and did not see the same potential in this technology. In addition to the political struggle with other product groups, lack of explicit support from management was another major barrier in this process. Rather than taking initiatives for own exploration and use of the technology, management mainly maintained a spectator role in this process, only issuing the needed approvals.

Also, at this time Microsoft NetMeeting was launched as a potential new product for Statoil, and the second author accepted the role of product manager for the development of this product (see Section 5.3.3 for presentation of this implementation project) to be able to participate in the further development and integration of meeting support technologies in Statoil. However, this new assignment temporarily shifted the focus away from GroupSystems.

Since 1998, the GroupSystems' meeting room and facilitation services at the company headquarters in Stavanger has been used regularly. There is also a permanent team of facilitators established, with responsibility for delivering services related to this meeting room. In 1999 a third meeting room was established in Bergen. Because of limited access to facilitators in Bergen and Trondheim, however, the use of the meeting rooms at these sites is more occasional. In September 1999, GroupSystems was also made available over the company network, enabling distributed meetings involving two or more linked meeting rooms as well as participants using GroupSystems from their office. These distributed meetings are usually also supported by audio, video and Microsoft NetMeeting in different combinations. Statoil IT rents the electronic meeting rooms and facilitators to other units in the company on an hourly basis. However, GroupSystems is still mostly used internally in the IT department, and the diffusion to other units in Statoil has been slow. In total, Statoil IT estimates that the company has conducted more than 300 meetings supported by GroupSystems since installation in 1996.

An interesting spin-off from the GroupSystems implementation has been the establishment of online websurveys as a new product area in Statoil. Despite only being marketed as an additional module in GroupSystems, this functionality quickly gained more widespread popularity than the main meeting support features themselves. Based on the successful use of GroupSystems websurveys, a new general online survey tool is now established as a new service from Statoil IT.

Barriers to Further Diffusion of GroupSystems in Statoil

The recruitment of facilitators constitutes the major bottleneck for further diffusion of GroupSystems in Statoil. Despite extensive internal training of facilitators (based on the vendor's training methodology) involving more than 100 participants, only a handful of these have continued to practise as facilitators. According to those responsible for the courses, the recruitment difficulty has much to do with the selection of participants for these courses. Treating this course as "any other" internal course, the selection of attendants has often been quite random, without these having any special motivation for learning facilitation skills. Besides motivation, becoming a facilitator may also be seen to require a special blend of personal characteristics. As expressed by one of the instructors:

> It requires a form of "call". You must have strong interests in this, and you need to have some specific personal characteristics. It is a big threshold. It is challenging in all ways: you both need to understand the tool use, you need to understand collaboration, you must be able to lead a group, it's a very challenging role to enter into, especially when you both need to master the technology, a rather advanced, universal tool with infinite possibilities that is quite hard to learn – and you need to know about methods for problem solving and collaboration. People have been scared off, I think.

The lack of available facilitators also resulted in the dismantling of a fourth meeting room that had been operating in Oslo for some time. The establishment of new meeting rooms also represents a challenge. Having already purchased a company licence, the software costs are not considered a barrier to further diffusion of the tool. However, the investments in new meeting room facilities and hardware, including audio and video conferencing equipment, are considerable. Finding a suitable room in itself is a challenge, as many of the meeting rooms in Statoil are "owned" by collocated work units.

The marketing of GroupSystems has also been limited, due to the restricted access to facilitators and meeting rooms. As the capacity for facilitation services has been fully deployed, there has not been a need for selling the technology to new customers. However, one of the informants thinks the internal marketing of the technology could have been better:

> I think we would have more requests for facilitation services if we had been better at selling and telling about the benefits from using it [GroupSystems]. We have not been explicit enough about what we deliver.

He thinks the marketing so far has had the character of "we've got this great tool, do you have any problems for us?", rather than focusing more on organisational needs and then introducing GroupSystems as a potential tool for solving these.

In the attempt to diffuse the technology outside Statoil IT, there have been several initiatives from Statoil IT of building alliances with other units that stand out as natural users of this technology. For example, there is a unit in Statoil called Change Support that facilitates processes for management, e.g. related to strategy development. However, this unit was not very receptive to the technology in the beginning – holding on to their traditional, manual facilitation techniques – and thus partly acted as a competitor to the services provided by the IT department. Though they have gradually gained interest in the technology and also conducted internal training courses, the diffusion has been slow with limited adoption so far.

In general, several of the managers in Statoil have also been reluctant to use GroupSystems. One of the facilitators reflects about this in the following way:

> I use the term "holy cow" about meetings, in the sense that some regard this as an arena that should not be "infected" with technology. Because this is a "free zone", where you can come to a meeting and drink coffee with the expectations from meeting participants often being very diffuse. This is my claim. And when you start to mess with technology and start talking about making the meetings more effective, then you get someone against you.

5.3.3 Implementation of Desktop Conferencing

The implementation of desktop conferencing in Statoil has occurred in parallel with the implementation of GroupSystems, both being related to Statoil's infrastructure for supporting meetings. In 1996, one of the key persons responsible for PC hardware in Statoil IT made an initiative for making Intel ProShare available as a product in Statoil for desktop conferencing (video and application sharing on PC). This product received D3 approval in 1997, and around 200 installations of ProShare with ISDN were established. However, these have not been used much, despite extensive marketing during the "I-days" in 1997. Several factors can be seen to have contributed to this. First, being based on ISDN this solution could not be run on the ordinary network, and also required special installations on each machine. As a result, the price for each installation was relatively high. Overall, the marketing of the solution was limited and also focused much on the audio and video functionality, although the video functionality did not prove to be that important. Finally, the desktop conferencing solution was regarded as competitive rather than complementary to GroupSystems as a meeting tool, although the functionality was different. The two solutions were marketed separately, without emphasis on the potential for combined use. Thus, tendencies of internal rivalry also hampered the uptake of this technology by the rest of the organisation.

During the summer of 1997, Microsoft NetMeeting was launched as an alternative product for desktop conferencing. Compared to ProShare, the NetMeeting solution had several advantages. It was a freeware product that could easily be distributed and run on the company LAN (but without audio and video), with lower costs per unit. Still, the functionality matched that of ProShare regarding application sharing, shared whiteboard, etc. After installation and testing, NetMeeting received D2 approval in December 1997. However, due to some shortages in personnel, the final preparation of the product with D3 approval was not made until May 1998. The solution was then made available for the company during summer 1998, with the second author as the product manager.

After this followed a period of internal marketing towards the units in the IT department responsible for delivery to other customers in Statoil and the related competence development. Early 1999 the marketing against customers in Statoil commenced and in March 1999 a large campaign for net meetings was launched, termed "Reisekutt '99" ("Travel cuts '99"), aimed at reducing the travel costs in the company. Using more than a man-year of resources, the campaign involved local training, demos and a comprehensive website on the Statoil intranet marketing the use of NetMeeting for reducing travel costs. In general, the team responsible for the campaign tried to mobilise the entire IT department in the marketing towards the different units. They also tried to "sneak in" some marketing of GroupSystems and Sarepta in this campaign, but without this getting much attention.

As a result of this campaign the team responsible for the implementation of NetMeeting was awarded the "I-prize '99" for the best IT application in Statoil, presented at the "I days '99" in April. At this time there was a strong focus on NetMeeting in Statoil, as a result of the pressure on reducing travel costs in general. A contributing factor here was the opening of the new Oslo airport at Gardermoen during autumn 1998, the location of which actually represented increased time for travel to Oslo. However, with the good financial results for Statoil the pressure for cost reduction can be seen to have decreased somewhat, and the use of NetMeeting has now flattened out at a lower level, with a relatively stable user group constituting most of the use. Thus, there still seems to be a large potential for further diffusion of this tool.

5.3.4 Integration of Meeting Support Technologies

During the first stages of the implementation of GroupSystems, this was still regarded as a competing product by the groups responsible for other collaboration technologies. This is illustrated by the fact that during the "I days '96" the team responsible for audio- and videoconferencing had a separate stand next to the GroupSystems stand, without any collaboration between these two.

To further the integration of the meeting support technologies, the second author took the initiative of forming a dedicated team for IT supported meetings. It took a year to establish collaboration and break down barriers between the different "technology camps". This finally culminated with a team that included representatives from both camps. The team was established in 1998, with the second author as the leader, and comprising both technical personnel responsible for maintaining and running the solutions, and those delivering facilitation (12 people in all).

a

b

c

Figure 5.2 Statoil's GroupSystems rooms in Stavanger, Bergen and Trondheim (as of 2000).

Having both technological and method/tool competence in the same team created synergy, and the team functioned as a combined management and delivery team.

This team upgraded all three GroupSystems rooms to include full integration of audio- and videoconferencing, NetMeeting and GroupSystems. Each GroupSystems room has a capacity of 12–15 participants, and is equipped with laptop PCs as workstations, audio- and videoconferencing equipment, and public screen projection (see Figure 5.2). The user friendliness of GroupSystems has also been improved through a fully automated login procedure.

As will be discussed in relation to the further strategies in Statoil, the integration efforts still continue. In 2000 the team for IT supported meetings was merged with the Sarepta team responsible for the general Lotus Notes/Domino applications.

5.4 Discussion of Experiences

5.4.1 Variation in Adoption Patterns

The description of the different implementation projects illustrates how the adoption and diffusion of these technologies have followed different trajectories. While Lotus Notes and NetMeeting have diffused relatively rapidly throughout the organisation, the assimilation of GroupSystems has been slower. Still, none of these technologies have yet been utilised to their full potential. For Lotus Notes there is also a large variation in the use of the different services. In 2000 a survey was conducted for charting the status related to the use of collaboration technologies in Statoil (see also Section 5.5.3). Based on system logs, Table 5.2 indicates the relative frequency of use for the different technologies.

This table shows that the Lotus Notes tools supporting asynchronous collaboration are still most used, while the use of tools supporting synchronous collaboration is yet more limited. Sarepta Archive was referred to in the survey as not being user friendly and having a high threshold for use, thus possibly explaining the relatively limited use of this tool. The table also shows how Sarepta Arena now has achieved widespread use, as a result of the dedicated support from the Sarepta team and more emphasis from management. The limited use of the Lotus Notes discussion databases can be partly explained by the fact that Sarepta Arena in practice is also used for this purpose. Still, the general use of discussion lists and news groups in Statoil also seems to be limited.

In general, it is also interesting to note how the barriers encountered in the adoption of these different technologies have varied. For Lotus Notes, the installation of the technology went relatively "smoothly", apart from some initial problems related to integration with existing technologies.

Table 5.2 Relative frequency of use of collaboration technologies in Statoil (as of 2000)

Application	Platform/tool	Relative frequency of use
Group calendar	Lotus Notes/Domino	Moderate
Meeting room reservation	Lotus Notes/Domino	Moderate
Sarepta Arena (document management, workflow)	Lotus Notes/Domino	Moderate
Sarepta Archive	Lotus Notes	Low
E-mail	Lotus Notes	Very high
News and bulletin boards	Lotus Notes/Domino	High
Web publishing (.homepage)	Lotus Notes/Domino	Moderate
Discussion databases	Lotus Notes/Domino	Low
Net meeting/data conferencing (MS NetMeeting + phone)	Microsoft	Low
Videomeetings	Tandberg	Low
GroupSystems	Paradox database, Windows client	Very low

This was a result of a well-planned implementation project with sufficient resources mobilised. However, the main challenge in the adoption of Lotus Notes has been that of changing the "collaboration behaviour" of the employees from focusing on individual production to sharing of information, to enable moving beyond the "automation" of existing routines. With few incentives established for this change, this has proved a difficult process.

Similar to Lotus Notes, the implementation of NetMeeting did benefit from a "felt need" in the organisation and a related focus and interest in the technology. However, with the main focus being on the Sarepta tools, the strategic emphasis on NetMeeting has not been sufficient for following up the successful marketing campaign. Thus the further marketing and sales process has not been strong enough to ensure further diffusion of the tool. Finally, the "culture" for travel in Statoil has also proved difficult to change, despite the proven economic benefits from net meetings.

In comparison with Lotus Notes and NetMeeting, the implementation of GroupSystems has been much more of an ad hoc nature. Although undergoing the product approval stages in Statoil IT, a formal implementation project for this technology was never initiated. Rather, it has been an evolutionary process driven by a small group of enthusiasts, and the lack of resources has marked the entire project. In general, the development of an electronic meeting infrastructure including IT supported meeting rooms and facilitation services has been a challenge. Especially, being able to recruit potential facilitators has proved a lot more demanding than just "ordering" personnel to undergo training.

Further, as compared to the other technologies the diffusion of GroupSystems outside the IT department has been more difficult. This may be partly explained by the more intangible nature of the potential benefits from this technology that may be less obvious to non-IT persons before actually having been exposed to it. Further, meetings in itself are a form of interaction/collaboration with long traditions and changing the nature and conduct of these has proved to be "touchy". Meetings are an important political arena and the inherent ability of EMS technology to eliminate power and status differences in a team may be considered less desirable by more dominating persons who rely on their position and verbal skills as their power base. Finally, the sometimes competitive relationship between the different collaboration technologies can be seen to have resulted in less attention and resources allocated to GroupSystems.

Still, it would be wrong to characterise the implementation of GroupSystems as a failure. There are several indicators of successful adoption and diffusion of the tool as well. The evaluations from the majority of participants have been positive, and after first having experienced the benefits from the technology many meeting owners return to the GroupSystems rooms for new projects. Further, the establishment of three permanent meeting rooms, an increased number of facilitators, and a broadened spectrum of customers (although still mainly within the IT department), should also be taken as indicators of success.

5.4.2 Organisational Influences

In addition to the different barriers discussed so far, it is also possible to identify more general causes for the problems related to the implementation of collaboration technologies in Statoil. These are more related to the organisation of the development of these services.

Lack of Standardisation and Coordination

The implementation of general IT tools for collaboration in Statoil has both been a result of strategic decisions and more or less planned initiatives from single units and people in the organisation. There has not been any established tradition in Statoil IT for continuous and systematic analysis of needs and possibilities related to new collaborative solutions. The more systematic systems development methods that traditionally have been the basis for delivering business specific solutions have not been applied to the same extent in the development of the general collaboration solutions. There are both technological and organisational reasons for this.

Through the implementation of Lotus Notes and Smartsuite as the common, basic tools for IT supported collaboration in Statoil in 1994–96, a complete application development platform was actually made available for all employees in Statoil. This represented a very low threshold for establishing new specialised solutions for processing semi-structured information, based on prototyping and evolutionary development. It did not take long before consequences of this became visible. The principle soon became: "new information type – new Lotus Notes database". This more or less random and evolutionary approach also in general characterised the development of collaborative solutions by Statoil IT, based on the Lotus platform and other technologies/products, despite the fact that Statoil IT at this time had a strong focus on standardisation and control of the product development process. The development took place in independent projects that were only under partial supervision from the central Product Council in Statoil IT. Central or overall strategies or plans for the development of collaborative solutions were not developed. At best, the management of this area was thus reactive and indirect. A consequence of this rather random and primarily technology-led development was that the need for services related to the implementation and use of new collaborative solutions to a little degree was addressed. The result of this was a large and relatively complex offering of poorly integrated collaboration tools, with a limited offer of related services. The restructuring of Statoil IT at this time, involving changes in processes and routines, may also have contributed to that it was long before these problems were uncovered and addressed.

Relationship with the Research Department

Another factor that further challenged Statoil IT in relation to this trend was that the research unit in Statoil during the 1990s had an active group for coordination and collaboration technology. This group gradually got a key role influencing the surveillance of new technologies and identification of potential new collaboration technologies and products, such as Lotus Notes. These researchers functioned both as customers and collaboration partners to Statoil IT in the realisation of new collaborative solutions, and in some areas they were also providing alternative technology services as competitors. The relationship with this research department resulted in a faster establishment of new collaborative solutions, but also in that Statoil IT was "forced" into a more evolutionary and unstructured form of work. In addition, the research department provided the projects with a more theoretically/academically oriented competence, through their collaboration with different external research environments, such as MIT Centre for Coordination Science and the Program for Applied Coordination Technology (PAKT) at the Norwegian University of Science and Technology. When Statoil IT also started recruiting personnel from the research unit and from the PAKT program, this inevitably resulted in a more "academic" focus for some of the new projects. Although the collaboration with the research unit should indicate that the development of new solutions was better founded on real needs in the organisation, in practice the technological possibilities and more general assumptions of the potential benefit of these were the primary basis for establishing new solutions.

5.5 Current Initiatives

5.5.1 The Next Step Project

During spring 1999 Statoil IT initiated a new product development program, termed *CO-2001*, with the intention of "doing something" with the current situation by providing a more complete offering of services and tools for IT-supported collaboration. A similar program was established for the further development of Statoil's IT infrastructure, named *Infra-2001*.

During fall 1999 the CO-2001 program got a push from a more visible strategic initiative in Statoil. To contribute to cost reduction in Statoil, the concern management decided to implement a so-called "Freeze NextStep" strategy related to the further development of the company's IT infrastructure and basic IT solutions. According to this strategy the existing PC hardware, printers, telephony and basic PC client software (including Lotus Notes and Lotus Smartsuite) was to be "frozen" until 2001, when a larger upgrade was to take place through the establishment of the "Next generation IT workplace". Then a new freeze period should take place. The two product development programs that still were in a start up phase were then merged into the *I-2001* project, which was to be the main vehicle for Statoil IT in implementing this strategy. I-2001 thus comprised the two former subprojects CO-2001, responsible for initiating the NextStep collaborative solutions, and Infra-2001, establishing the NextStep IT infrastructure.

The I-2001 project was initiated by a pre-study where eight selected vendors were invited to present their future visions and concepts for Statoil's NextStep IT workplace. Based on an evaluation of the vendors' presentations of their concepts, two vendors were selected for further collaboration. Not surprising, these vendors were Microsoft and IBM/Lotus. This was followed by a "technical feasibility study" where the products from these two vendors were assessed according to the possibilities they offered for implementing the NextStep IT workplace. The study was conducted through separate collaborative projects with the two vendors. The aim was to develop an architecture for the NextStep IT infrastructure and collaborative solutions based on the vendors' standard products, minimising the use of tailored solutions for Statoil.

In parallel with this work an "organisational feasibility study" was conducted, mapping the company's use of existing collaboration solutions and future challenges related to collaboration and collaborative tools. Based on these results the project did a first attempt during summer 2000 of developing an overall strategy for the further development of services and tools for e-collaboration in Statoil. This strategy was intended to serve as the basis for the further development of the IT department's services and general tools for e-collaboration, including the NextStep collaboration solutions. The I-2001 project should deliver the strategy at the request of the IT unit responsible for the existing tools and services.

The main result after the I-2001 project had run for one year was that Microsoft Windows 2000 was chosen as the platform for the NextStep IT infrastructure and Lotus Notes/Domino R5 was decided to be maintained as the platform for the general collaborative tools. In addition, Microsoft Office 2000 was chosen as the new standard office support tool for the company, replacing Lotus Smartsuite. The strategy suggested by the CO-2001 project was never formally approved by the management in the IT department, but was still used as the initial basis for the further development of the NextStep collaborative solutions.

The establishment of the NextStep IT workplace started in October 2000. The project had then changed its name from I-2001 to NextStep, and a third subproject was established for supporting the implementation and training related to the NextStep IT workplace. The main focus of this subproject, however, was on the technical aspects of the implementation, and the deliverance of sufficient training in each separate tool to enable the users to get "up and running" in the new IT workplace with the minimum of disruption of their work. The new IT workplace was made available for all employees in Statoil during the summer of 2001. The plans for the development of the collaboration tools are described in more detail in the following sections.

5.5.2 Related Projects

E-collaboration Team

In parallel with the establishment and conduct of CO-2001 and I-2001, a further development and integration of the IT department's services and collaborative solutions were also taking place. During spring 2000 this culminated with the merger of the Sarepta team and the team for IT-supported meetings into a central team for "e-collaboration", responsible for delivering support to consulting, deployment, adoption and facilitation of e-collaboration in Statoil. The main deliverance from the team is introduction and support for the existing standard solutions for workflow and document management (Sarepta) based on Statoil's portfolio of internally developed Lotus Notes/Domino standard tools. In addition, the portfolio includes deliverance of standard tools for net meetings and facilitation of e-collaboration using GroupSystems. During autumn 2000 the responsibility for this team was transferred from Statoil IT to the staff function in Statoil with responsibility for change support.

Projects Outside the IT Department

CO-2001 originally aimed towards a more integrated and coordinated further development of the general collaboration tools, but only to a certain extent managed to include the existing product development projects. Several other projects for further development of general collaborative solutions continued independently of I-2001/CO-2001, especially in relation to establishing general solutions for publishing on the intra- and extranet, and web-based solutions for information sharing and collaboration with external partners. These projects were to a lesser degree centrally managed by the IT department, and to a larger extent driven by user groups outside Statoil IT.

5.5.3 Needs Analyses

The product development program CO-2001 and later the I-2001 project gave the IT department an opportunity to conduct somewhat more systematic and complete needs analyses than before. The most comprehensive of these was conducted during spring 2000. Interviews, workshops and a major survey were used for charting and analysing the use of existing general collaborative solutions and future challenges related to these. The result that was focused in the report from this survey was the state in relation to training of the organisation in use of the collaborative solutions. Very few of the participants in the survey stated that they had undergone any particular degree of training in using the tools. The need for services related to improved training, application support and realisation of benefits was thus clearly confirmed by this survey. The users' assessment of the single tools also confirmed that individual tools such as Lotus Smartsuite and Lotus Notes e-mail were in reality still dominating the market for production and exchange of information. These tools were both clearly more used and more popular than "open" common tools such as group calendars and Lotus Notes databases. Meeting support systems such as Microsoft NetMeeting, GroupSystems and IT supported meeting rooms were also used relatively infrequently, and only achieved a modest assessment regarding popularity. Related to future needs and challenges, "boundaryless collaboration" and improved information management were emphasised. The needs for simplification, user friendliness and improved integration between the solutions were also clearly stated.

The studies conducted by the NextStep project did, however, only to a small extent focus on future scenarios and new technological possibilities, and this also needs to be taken into account when interpreting these results. Apart from the focus on training, these studies did not go into other possible causes for this state, in relation to other organisational factors such as management, organisation and collaborative culture and attitudes. Still, they do indicate the stage Statoil has reached regarding support and improvement of collaboration through the use of general IT solutions. Sharing and management of individually produced information ("information and knowledge management") still gains focus, both regarding evaluation of established solutions and analysis of future needs and challenges. Support for knowledge creation in groups is less focused. In general, the production tools are emphasised more than tools for coordination and decision support for groups.

5.6 Towards an Integrated Collaborative Solution

One of the first results of the establishment of the CO-2001 subproject in 1999 was the development of a concept for IT-supported collaboration. This was intended to give a holistic understanding of what collaborative environments and processes generally consist of, and how IT tools may support these. This concept was used throughout the CO-2001 project, as a vehicle for analysis of the existing tool portfolio, needs analysis, feasibility study, strategy development and later on also the detailed planning of which collaborative solutions were to be included in the NextStep IT workplace in Statoil. The concept was based on former work done by participants in the project and the unit in the Research Centre, and represented a pragmatic approach rather than a "scientific" work. According to this concept a collaborative tool represents a certain combination of and access to generic *functions (methods)*, *communication styles* and *information types*. The functions consist of the three main functional areas of *coordination*, *production* and *decision support*, which again are decomposed into a set of generic sub-functions. A communication style describes how a function is available with respect to the actors' presence in time and space, use of media and authentication. An information type describes a type of information managed by the tool. The information types in practice distinguish "general" collaborative tools from "special"

Table 5.3 NextStep Collaborative Solutions in Statoil

Functional module	Functions
Personal workspace	Web-based "Business to Employee" (B2E) portal providing simplified access to general solutions, business specific solutions and information resources
Unified messaging	One mailbox for all message types (e-mail, fax, voicemail, SMS, etc.)
Calendar and scheduling	Group calendar, meeting scheduling, resource reservation
Conferencing	Meeting coordination, screen sharing, whiteboard, chat/ instant messaging, presence/awareness, audio, video
Virtual Collaboration Room	Project and activity structuring, structured co-authoring of all general information types, meeting and decision support, presence-status (awareness) on information and users
Archive	Statutory archive and experience archive
Search	Information search and management (content management)
Authoring	Individual authoring tools for rich text, numeric, presentation, drawing/graphics, audio, video
Collaboration facilities	Improved common IT-supported meeting rooms equipped for multi-participant use of Conferencing tools and Virtual Collaboration Room

ones in a given organisational context. General collaborative tools are used on general information types occurring everywhere in the organisation.

However, when the final choices of general collaborative solutions for the NextStep workplace should be made and approved during fall 2000, the generic functional model in the collaboration concept was abandoned for an even more pragmatic approach to describing the planned total solution. The generic functions in the concept were then mapped into a few main functional modules, as shown in Table 5.3. These formed the conceptual basis for the NextStep collaborative solutions in Statoil. Several details are not shown in the table (as, for example, the degree to which the functions are planned to be made available to external actors), but it provides a picture of what is being focused in the solution.

The initial aim was to establish a solution with all of the functional modules described in Table 5.3 by 2002. Some of the new basic functions were established in 2001, at the same time as new PC-clients and new PC-infrastructure based on Microsoft Windows 2000 and Lotus Notes R5 were introduced. The general development trend emphasised for the NextStep collaborative solutions comprised web interface, multimedia support and use of commercial tools "out of the box" as much as possible. Table 5.4 shows the products chosen for the 2001 delivery.

Table 5.4 Functions and products in the 2001 deliverance in the NextStep project

Functional module	Products
Personal workspace	Lotus K-Station 1.0
Calendar and scheduling	Lotus Notes R5 Calendar
Data conferencing with external parties	Lotus Sametime 2.0
Search	Lotus Domain Search and Inktomi Search
Authoring/office support	Microsoft Office 2000

The 2001 deliverance by the NextStep project was completed according to plan, and described as a success by the organisation. As of August 2001 all of the solutions in Table 5.4 were in normal operation. The planning of the continuation of the NextStep project started in April 2001. The initial functional model (shown in Table 5.3) was used as the basis for this planning, gradually being expanded and detailed related to specific functional areas. Among the key focal areas in the further development of an integrated collaborative solution are:

- improvement of mobile services (mobile phone and PDA);
- e-Learning;
- information management (archiving, search and content management);
- workflow and decision support;
- improvement and expansion of the portal solutions and the solutions for external collaboration.

Meeting support systems are also relevant in this, but will most likely get less emphasis as isolated solutions in the future. Rather, these are expected to become better integrated with the other solutions.

5.7 Further Challenges

The current initiatives taking place in Statoil represent ambitions for further development of general IT tools and services for collaboration. The NextStep project and the upcoming projects in the IT department contribute more and new enabling general collaboration tools, and there is increasing acceptance of how emphasis on services related to implementation is necessary for extracting the maximum benefit from the tools.

One of the challenges for Statoil and the IT department is to figure out at which level and how the implementation services are to be delivered, both internally in Statoil and in relation to collaboration with external parties. With the merger of the two former support teams into the new e-collaboration team, an important step has been taken towards the further integration of the services. However, the composition of this new team is still somewhat biased in favour of the asynchronous tools, thus resulting in less emphasis and development of the services related to synchronous collaboration (including meeting support).

The existing solutions mainly support asynchronous production and sharing of information, and to some extent coordination of collaboration. These processes will be further improved and simplified through current projects. Related to the aim of supporting total collaborative processes, support for group decision processes and integration/flow between coordination, production and decision processes remains to be provided, in addition to support for synchronous conduct of these processes (in meetings). The latter requires facilitation support, a type of service and competence still scarce in Statoil. To establish acceptance for the use of IT support in meetings is by itself a challenge. The benefits are obvious, but the skills in leading and facilitating meetings are equally varied. This relates both to co-located and distributed meetings (net meetings). IT supported meetings enable new work practices that can be seen to challenge the existing culture related to both meeting and travel in Statoil. However, the emergence of meeting technologies is continuously being challenged by technological and societal trends that to an increasing degree emphasise the individualistic, competitive IT user.

Support for asynchronous processing of information and knowledge that has already been developed is being emphasised, while actually supporting the development of information and knowledge in groups (knowledge creation) is addressed to a lesser extent. This makes it generally difficult to introduce IT solutions that support synergy-creating collaboration in teams. In general, technology trends will always have greater impact on strategic technological choices than integrative perspectives focusing on holistic thinking and collaboration concepts. However, we believe it is important to challenge these trends for being able to support integrated collaboration processes. Related to this, measures for increasing the general level of competence in communication and collaboration are important.

The challenge will be to contribute both to a heightened competence related to practising collaboration in general and to provide the necessary services to support the adoption of IT tools for collaboration. Problems related to the technology are expected to decrease, both related to internal collaboration in the company and collaboration with external partners. In a few years' time, all collaboration tools known today will be available "anywhere on any device", but multi-participant collaboration will still require facilitation that can only partially be automated and supported by the tools. Access to tools for asynchronous coordination and production in teams will no longer constitute a competitive advantage by itself. To advance further, the focus needs to be shifted towards the integration, management and facilitation of the collaborative processes.

5.8 Lessons Learned from the Statoil Case

Statoil represents an interesting case in that it offers the possibility to contrast and compare the experiences from implementation of a range of collaboration technologies within one organisation. As discussed in Chapter 3, previous case studies have mainly reported experiences from the implementation of single technologies, without addressing the broader technology portfolio of which these technologies are part. Similar to previous studies (e.g. Karsten and Jones, 1998; Orlikowski, 1996a,b), the organisational adoption and diffusion of the different collaboration technologies in Statoil has been a gradual and relatively slow process, spanning several years for each technology. Several barriers have been encountered in this process, both related to the initial adoption and the further utilisation of the technologies. As can be expected, the major challenges have been related to gaining organisational acceptance and changing the collaborative behaviour of the employees. However, the technical and infrastructural challenges in the implementation should not be underestimated.

Adequate allocation of personnel and resources for dedicated follow up of the implementation projects has proved instrumental for success. In all instances where these conditions have not been present, this has had a direct impact on the progress in the projects regarding adoption and diffusion of the solutions. The importance of enthusiasts and champions in driving these projects also tends to make them highly dependent on single individuals, with resulting vulnerability regarding capacity and continuity. The Statoil case also illustrates the importance of integrating the implementation efforts for related technologies. Only after the merger of the support groups for the different technologies has it become possible to achieve a common focus for the further development. However, to obtain an equal balance in the deliverance of the different technologies will require more resources to this team, and a strengthening of the focus on synchronous collaboration technologies. Further, although the CO-2001 and NextStep projects signal a strong focus in Statoil on the potential represented by collaboration technologies, there still exists a need for more explicit management focus and support to be able to advance further in the process of establishing effective utilisation of the technologies.

The process related to deployment and adoption of collaboration technologies in Statoil still continues, and it is therefore not possible to present any general conclusion on the degree of success in their efforts. With the strengthened focus on the Internet and web, new collaboration tools and applications become available that need to be tested and evaluated for possible adaptation to the organisation. Similar to many companies today, Statoil is still at a relatively early stage of applying collaboration technology for knowledge creation and management, rather than merely as a vehicle for boosting individual productivity. The recent user survey indicates that much work is left regarding training of the employees in effective use of the collaboration technologies, and also for raising the general competence level in the organisation needed to exploit the potential of the technologies. If Statoil is to become a leading company in practising synergistic IT-supported collaboration, a stronger focus is needed on the collaboration environments and processes and how these can be improved through several means, including technology and facilitation in a more integrative perspective.

6

Implementation and Use of Collaboration Technologies in a Multinational Engineering Group: The Case of Kværner

Bjørn Erik Munkvold

6.1 Introduction

This case study presents the experiences from implementation and use of collaboration technologies in Kværner, a multinational engineering group. Kværner has implemented a global area network for supporting communication and information exchange among the companies in the group, serving as the backbone for use of other collaboration technologies such as document management and desktop conferencing. The case study presents the lessons learned from the implementation of this network, and the challenges related to achieving full utilisation of the collaboration technologies in Kværner. The chapter is based on interviews with representatives from central IT management in Kværner, and three Kværner companies (two in Australia and one in Norway). Additional information was collected from the Kværner website (www.kvaerner.com).

6.2 The Kværner Group

Kværner is a Norwegian registered engineering and construction group with historic roots back to 1853. In 1996 Kværner acquired the UK based Trafalgar House engineering group. With a price of approximately US$1.2 billion, this was the largest take-over ever by a Norwegian firm and made Kværner among the largest engineering groups in the world.

As part of this acquisition, Kværner moved its administrative headquarters from Oslo to London. In the years following, Kværner experienced severe decline in several of their markets, especially in the Asian region, thus proving their offensive expansion strategy had been too optimistic. As a result, a major restructuring of the company was initiated in 1999, including change of top management. Kværner is now focusing its activities around two core business areas, engineering and construction (E&C) and oil and gas (O&G), with other current activities (including shipbuilding and pulping equipment) being sold. The Kværner group has annual revenues of approximately US$8.5 billion, with some 35,000 permanent staff located in almost 35 countries throughout Europe, Asia and the Americas.

6.3 Establishing a Global Collaborative Infrastructure: The KINET Project

In 1995, Kværner signed a three-year contract with Telenor (former Norwegian Telecom) as their vendor partner for developing KIGAN (Kværner Internal Global Area Network), with British Telecom and MCI serving as subcontractors. The contract included the establishment of telecommunications services for voice, e-mail and data worldwide in the Kværner group. The technology chosen as the basis for the network was frame relay, a WAN (wide area network) technology providing permanent virtual circuits for transmission of e-mail and attached documents. After the acquisition of Trafalgar House in 1996 the name of the project was changed to KINET (Kværner Integrated NETwork). Figure 6.1 shows the topology of KINET, comprising four regional hubs in Oslo, London, Atlanta and Singapore. The individual Kværner offices are connected to KINET through the respective regional hubs.

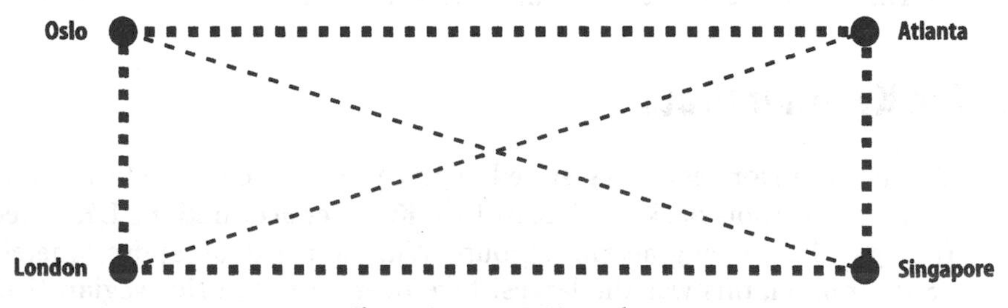

Figure 6.1 KINET topology.

6.3.1 Aim and Scope

The KINET project originally started out as an effort to reduce the communications costs in Kværner. Due to the decentralised structure of the Kværner group there was little coordination of communication needs and related technology use among the companies, and the top management saw great potential savings from being able to negotiate a common contract for Kværner with a single telecommunications vendor. At the same time the number of projects involving collaboration between two or more Kværner companies were increasing, and there was a need for more effective ways for communication and information exchange. Further, KINET was seen as an important vehicle for supporting the increasing internationalisation of Kværner's operations. The establishment of this type of network was also considered to have a marketing value. Especially in the oil and gas sector, not having this type of network could actually represent a major drawback, as electronic communication was becoming a prerequisite for pre-qualification for tenders with all related information being distributed electronically. Several of Kværner's competitors were also undertaking similar networking projects. Thus, the goals of the KINET project gradually developed from merely focusing on cost reduction to also encompass more strategic aspects. KINET should enable all companies and employees in Kværner to communicate electronically, regardless of business area or geographic location. By this, Kværner envisioned the following benefits:

- more efficient collaboration between companies in Kværner;
- development of "Centres of Competence" that could be accessed by other companies in Kværner through the network;
- increased coordination between the companies in the Kværner group, enabling large scale benefits in purchasing, service contracts, etc. with third party companies.

With KINET as the network infrastructure, Kværner signed a contract with Digital and Microsoft in 1996, for developing an intranet solution for the entire group.

6.3.2 Organisation of the Implementation Project

The implementation of KINET was contracted out to Kværner Engineering in Oslo, with the core implementation team consisting of five people with mainly technical and engineering backgrounds. Four regional centres were established in Houston, London, Singapore and Sydney, handling all contracts in the respective areas and acting as a focal point of contact for Kværner. In addition, Kværner established a "KINET Advisory Board" consisting of top managers in Kværner, including two of the eight vice presidents. This board had internal meetings every third month, and meetings with Telenor twice a year. There was extensive collaboration between Kværner and Telenor, with regular meetings both on management and operational level. A core team was established in Telenor consisting of an account manager, service manager and design manager. A special service concept was developed, specifying how the deliverance was to be handled, including quality assurance.

6.3.3 The Adoption Process

Although initiated by the top management in Kværner, the adoption of KINET was fully decentralised with each company being free to decide whether to become part of the network. The costs for each company were calculated as a dividend of the total annual costs of the Telenor services, based on the size of the company. In addition came any investments in servers and clients, as well as licences for the e-mail system. The internal marketing of the project was handled entirely by the KINET implementation team, by visiting each company and trying to convince the management to adopt the system. To ensure sufficient mandate, the implementation team made it their policy only to deal with the companies' top management, either the CEO or Chief Financial Officer, although with the IT manager usually present. Companies deciding to adopt the system signed a contract directly with Telenor, who handled all installations. Apart from this, the companies only dealt with the KINET operation. The four regional centres supported the adoption process in their respective regions. Apart from this, there were no particular project "champions" outside the implementation team.

The original plan was to have all companies linked to KINET by the end of summer 1996, but there were several delays in the project. More than a year later all companies were finally linked to KINET. However, the network traffic increased rapidly, and during 1997 KINET reached the status of being regarded as "mission critical" for Kværner.

6.3.4 Barriers to Implementation

The implementation team encountered several barriers in the implementation process, both related to the adoption of the technology in the different companies and technological issues. Table 6.1 provides an overview of the different issues identified.

Barriers in the Decentralised Adoption Process

The decentralised decision structure in Kværner made the implementation process slower and more resource demanding than the implementa-

Table 6.1 Barriers encountered in the KINET implementation

Barriers related to the decentralised adoption process
- adoption costs
- lack of incentives for adopting the technology
- relations to external parties
- cultural factors
- limited capacity of the implementation team
- resistance from local IT departments

Technological barriers
- heterogeneous infrastructure
- incompatibility with existing solutions
- problems with performance and stability

tion team had expected. They had to invest much time and effort into marketing KINET in the different companies and persuade them to adopt the system. This process was described by the project coordinator as "generating a lot of friction heat". A combination of the issues listed in Table 6.1 can be seen to have contributed to the inertia in this process.

Adoption costs

In their marketing of KINET, the implementation team documented considerable potential savings in communications costs by adopting the technology. For example, their cost-benefit analysis showed savings on the voice part of the service that alone would justify the investments in the technology. Still, several of the companies rejected the technology due to the costs being too high. According to the project coordinator, many of the companies were seemingly unaware of the actual costs from existing communications media such as telephone or fax. However, the companies' initial concerns about the adoption costs also need to be seen as related to the other issues listed in Table 6.1, such as lack of incentives for adopting the technology and existing relations to external parties.

Several of the companies chose to "sit on the fence" and wait for others to adopt the technology first. As expressed by the project coordinator, "establishing a critical mass when everyone is waiting for each other is impossible". To break out of this deadlock, the implementation team talked the Kværner administration into financing the establishment of regional access nodes, to make the process of connecting to the network easier and less expensive.

Lack of incentives for adopting the technology

Several companies did not see a need for the services provided by KINET. The real benefits from adopting this technology were to follow from more efficient collaboration with other organisations in the group. Despite the trend towards more collaborative projects, however, for several companies the existing level of collaboration with other Kværner companies was low. Companies not yet involved in this type of collaboration thus had difficulties relating to these potential benefits, being content with their current situation. Some companies were already using alternative services based on the Internet and CompuServe. Regarding the information they transmitted as insensitive, these companies did not require high network security and were therefore content with the existing solution.

For many organisations, the basic ideas of the KINET project involving increased coordination and collaboration among the Kværner companies represented a new way of working. The implementation team therefore also had to take on the role as missionaries for these ideas. A barrier to this was the lack of coordination between the management of the different companies, resulting in companies building up in-house expertise and resources instead of exchanging this with other Kværner companies. Second, in an industrial company like Kværner, IT is given lower priority than other engineering and production technologies, and many companies were mainly focusing on their short-term needs regarding communications technologies. Finally, although the KINET project was backed by the Kværner concern management, there were no management "imperatives" or explicit signals expressing the management's wish for adoption and use of KINET. According to the project coordinator this would have helped in the adoption of the technology.

Relations to external parties

The Kværner companies typically had their individual networks of external collaboration partners, including local vendors and subcontractors. The political benefits of maintaining existing relations with third parties in the local market and community would often outweigh the potential benefits from initiating collaboration with remote companies in the Kværner group. This also limited the perceived need for communication with other Kværner companies. However, there were also examples of how the relations to external parties actually were a driving force for adopting the technology. In one company, the possibility for communicating more effectively with their main customer was given more weight in the decision to adopt the technology, than communication with other Kværner companies.

Existing bonds to vendors were also a potential barrier in the adoption process. For example, in the UK several of the companies had problems with accepting British Telecom as their supplier, due to their relations with other telecommunications providers.

Cultural factors

Cultural differences also added to the complexity in the decentralised adoption of KINET, and the implementation team experienced how the presentation and marketing of the technology had to be adapted to the local culture of the companies. Issues considered to be straightforward in one organisation could be regarded as problematic in another. In general, developing a common culture in a distributed organisation like Kværner was regarded as extremely difficult, if not impossible. This was also due to the fact that the majority of the Kværner companies had been acquired, often without even changing the management. Still, the potential for KINET to contribute to the development of a more common culture in Kværner was used to some extent in the marketing of the network.

There were also examples of how protective policies enacted by national telecommunications companies in the Asian region caused hindrances in the project. In China, Singapore and Indonesia, the local telecommunications companies were protecting their local markets against foreign companies, by imposing strict regulations that caused delays for British Telecom serving as the subcontractor. For example, in Singapore British Telecom was only allowed to have its own network node placed in the Singapore Telecom building. And the Kværner operation in Jakarta was told that it would take over a year to get a permanent line established. Luckily, they found that there were already cables installed in the walls of their office building.

The acquisition of Trafalgar House also implied the need for integrating different company cultures, with related political struggles for positions in the new organisation. The extent to which this constituted a challenge varied with the degree of overlap between the functional areas of the two groups' companies in the different regions.

Limited capacity of the implementation team

Compared to the size and complexity of the KINET project, the core implementation team was very small. The high pressure on the team frequently resulted in capacity problems, with companies becoming impatient. When one of the key persons in the vendor organisation became ill, this affected the pace of the entire project. The project was therefore vulnerable to key person dependencies. The project coordinator considered putting more people on the team, but had problems with finding persons in Kværner with the right competence for this project, i.e. a combination of broad project experience and competence on WAN technology. He was also reluctant to use external recruitment, due to the need for training in Kværner's operations. As a consequence of the limited capacity of the implementation team, a detailed progress plan was not made until some time into the project, resulting in a somewhat ad hoc based operation mode for the team.

Resistance from local IT departments

In some companies the IT departments were sceptical about KINET, thus also influencing the adoption decision process in their company. The project coordinator partly ascribed this to the "not invented here" syndrome, i.e. resistance grounded in the fact that they had not taken part in the planning and development of the technological solution.

Technological Barriers

There were several technological barriers in the process, mainly resulting from the diversity in existing technological infrastructure in the Kværner companies. In addition, the merger with Trafalgar House also involved the integration of their existing wide area network with KINET.

Heterogeneous infrastructure

The Kværner group is an extremely heterogeneous environment regarding size and technical infrastructure of the organisations, ranging from companies with thousands of employees to small offices with only a handful of people. According to the project coordinator, a serious mistake in the project was the decision to choose one particular technology, instead of building a network architecture that would serve the needs of all the different companies. A survey among the organisations was conducted in an early phase of the project, to map their technical infrastructure and communication needs. However, this did not give as much information as needed, since many of the organisations were not able to specify their future needs in this area. As a result, the implementation team had to spend much time and energy on finding ways for the small, "far away" offices and mobile employees to be able to access the network. This has been one of the major reasons for the delays in the project. Due to limited experience with similar projects and lack of local knowledge in some of the regions, the vendor was unable to provide much support in this process, and the implementation team had to undertake most of this work themselves.

Incompatibility with existing technologies

Incompatibility between existing technologies in the companies and the KINET technology also represented a barrier in the adoption process. "Every" existing e-mail program was being used in the different companies, and there were several problems with integrating these. For example, problems with exchanging address information between Lotus Notes and MS mail in KINET was stated as a major reason for the slow adoption of the system in one of the Kværner companies. The same company also expressed discontent with the standardisation of e-mail systems as part of the KINET implementation.

Problems with performance and stability

During the initial stages of the project, there were problems with stability and performance of the e-mail services in KINET. As expressed by the design manager in one of the companies, these problems could easily influence the adoption of the technology:

> The system has to be stable and 100% reliable, so that you can be sure that your e-mail gets through. And if something goes wrong, it is crucial that the system provides status messages about this. If this is not the case, companies will not dare to use the system, in fear of that their e-mail does not get through.

6.4 Kværner's Use of Collaboration Technologies

The KINET today constitutes the backbone for the use of collaboration technologies in Kværner, providing e-mail and Internet/intranet services as well as serving as the platform for distributed use of different collaborative applications supporting the engineering disciplines. Due to the large variation in projects and needs for technology support for the different Kværner companies, it has not been regarded as feasible to implement any standard engineering and design applications at the overall group level. However, such standards exist within each division. The following section provides a brief overview of the current organisation of IT services in the group. This is followed by an overview of the different technologies in use, including experiences and problems in their deployment.

6.4.1 Organisation of IT Services in Kværner

In 1996 the Group IT function (GRIT) was established, with responsibility for setting high level standards such as NT domain structure, management of KINET, and e-mail and Internet standards. Prior to this no coordinating body for the use of IT services in Kværner existed, but the acquisition of Trafalgar House along with the increasing globalisation of the operations pushed forward a need for this. GRIT was mainly staffed by IT personnel from the Kværner headquarters in Oslo. At the beginning of 2000 the functions maintained by GRIT were outsourced to Equant, an international telecommunications provider. Equant has established a dedicated operation for maintaining KINET, also incorporating some of the former personnel from GRIT. Equant is also responsible for the further strategic development of the KINET services.

This outsourcing is part of a general trend in Kværner of downsizing the IT function, changing its role from a large independent unit to smaller service providers for the disciplines. The role of the regional IT offices is now mainly to provide the infrastructure for the engineering disciplines in the form of access to the KINET backbone, a standard PC on every desk, access to the standard applications (within each discipline), and the necessary "housekeeping" related to this (backups, etc.). The competence on how to use the different applications is now located within the different disciplines and each discipline is responsible for establishing routines for best practice in the use of the technologies. Further, all software development related to the proprietary engineering applications developed in Kværner has been sold out, with the applications now being marketed as commercial packages. In general, IT is no longer considered a specialist tool, and the IT function has become a support service similar to personnel and finance rather than a major department in its own right.

Both divisions in Kværner use KINET and the common NT structure established by the former headquarters in Oslo, but apart from this there are no standards set for applications and systems across the two divisions. Each division applies a set of standard, proprietary systems for supporting their engineering projects, and there are also standards for hardware and configuration. This means that within each division, employees can use their portable computers and log on to the network from any Kværner office in the world. There is also work under way for integrating further the technologies used in the two divisions.

6.4.2 Overview of Collaboration Technologies in Use

As discussed in Chapter 2, it can be difficult to define the boundaries for what is to be regarded as collaboration technologies. Table 6.2 lists the technologies that may be considered relevant in the Kværner case.

KINET can be considered the major collaboration technology in Kværner, providing the group-wide e-mail and intranet services, as well as the infrastructure for distributed use of other collaborative applica-

Table 6.2 Collaboration technologies used in Kværner

KINET
- E-mail
- Intranet
Engineering applications
- Project management tools
- Integrated engineering data repository
- CAD applications
- Enterprise Document Management System
Desktop Conferencing/Application sharing
Videoconferencing

tions. E-mail is the most widely used tool, for coordination and information exchange. The intranet comprises all types of internal company information, including information department services (news, press releases, annual reports, periodicals, etc.), company profiles, common contracts and purchasing experiences, "who knows what in Kværner" and "jobs in Kværner". The Oil and Gas division has also established discussion groups as part of their intranet services. The maintenance of the intranet is decentralised to the different disciplines and administrative functions, with web editors appointed in each area. The intranet services have also been used as a means for stimulating adoption and use of KINET, by providing various company information exclusively through this communication channel, and thus "force" companies into using the system.

Within each division, there is a standardised suite of proprietary systems being used for supporting the engineering projects. Building on the services applied in Trafalgar House, the E&C division is using a project management tool, an integrated engineering data repository, CAD applications and an enterprise document management system (EDMS) based on Documentum. The O&G division is running a suite of parallel systems, covering instrument design, electrical design, procurement, materials management, etc. O&G uses DocuLive as their document management system. Distributed use of these systems enables Kværner to move its operations between several locations during different stages of a project, channelling the detail engineering to centres in countries with high expertise and relatively low cost. Kværner is marketing this concept as "high-value, low cost engineering" (Cockshoot, 2000). This involves collaboration with operations in India or the Philippines or with third-party organisations in the Far East, Middle East, North Africa and Latin America.

The use of the EDMS for creating project archives is of key importance to this work concept. The management of documents in engineering projects is a complex task, with the number of documents being produced during the design and build phases of larger projects exceeding one million. When running distributed projects, a document management server is normally placed at each site, with the documents in these being synchronised through replication over night. The document management system is also used as a repository for internal documentation in Kværner such as engineering procedures. The EDMS also includes workflow management functionality, enabling pre-programmed distribution of documents for review and approval.

These engineering applications form the basis for Kværner's business-to-business (B2B) strategy. Based on a virtual private network, Kværner has established a service termed ECxtra, that enables secure communications and access to project documents in Documentum over the Internet (White, 2000). Through this, engineering data and project documentation are increasingly shared with external parties such as clients, subcontractors and vendors.

Kværner also makes increasing use of Microsoft NetMeeting for desktop conferencing and application sharing. Use of this tool has been especially successful for training. Instead of having people travel to different sites for attending training, the application sharing features in NetMeeting are used to talk people through the use of new databases, etc. Similarly, the technology is also used for troubleshooting. The only difficulty here is the time zones, with 11 hours time difference between the UK and Australian offices. NetMeeting has also been used in combination with multi-part videoconferencing, for distributing presentation slides to the different sites. Typically, they have then used two screens, one being the videoconference screen displaying the pictures from the different sites (e.g. split into four in a five-party conference), and the other being the NetMeeting screen. However, unless the slides have been very simple, there has tended to be problems with the time taken to refresh the slides at each site. Setting up NetMeeting across four or five different offices has also been a challenge. Thus, this use of the tool is considered less effective:

> We set up the videoconference, and then we spend the first half an hour getting everybody's NetMeeting synchronised. It's a big waste of time. It's much easier to send the slides out in advance to everybody, and the person who is making the presentation can just say "next slide", and everyone locally presses the button and gets the slide instantaneously, rather than building it over a minute or so for a complicated slide. And then you wait for a minute, it's pointless. NetMeeting is a waste, too complicated for this. There is no need for NetMeeting in that instance.

In addition to these technologies Kværner uses videoconferencing for project related and administrative meetings. They also see this as an important medium for supporting the development of interpersonal relationships in their distributed engineering teams. However, effective use of this medium usually requires that initial face-to-face introductions have been made and some social interactions have taken place. In general, Kværner still puts emphasis on face-to-face interaction, and invests resources in teambuilding efforts at the outset of projects.

> Actually that's the oldest collaboration technique of all, and something that we use quite a lot is when we set up a project team we physically move the people to sit together, all the people working on the project, whether they be instrument engineers or electrical or civil or process or whatever, we tend to put them all in one physical area to collaborate.

6.4.3 A Case Example: The Laminaria Project

An example of advanced use of the collaboration technologies for supporting distributed project operations is the production of the topsides for the Laminaria production ship developed for Woodside Petroleum in Australia. This was a large, detailed design project, involving several hundred thousands of man-hours. Starting off in Oslo, after 16 months the whole project moved to Perth, Australia, at about 70 per cent through the detailed design. The project used the full suite of IT systems in the Oil and Gas division. Using a combination of KINET and project dedicated links, the project established real time connection between the large number of physical sites, including Oslo, Monaco, Abu Dhabi, and later Korea, Singapore, Perth and Fremantle. The major collaborative tool used in the project was the document management system (DocuLive), comprising the full project archive for all technical documents, vendor documents, correspondence between parties, as well as all meeting minutes and project internal documents. Electronic workflow management was also used within this product, and documents issued for review were put into electronic folders and sent to various distribution routes with approval and comments retrieved electronically. Throughout the project, engineers in Oslo, Perth and Singapore logged into the engineering databases.

Use of the systems was also integrated with external parties. All systems were provided to the client's team, and they could read any of the documents as they were in production. Although they did not have write access it became almost an integrated team. The major fabricators in Singapore and Korea were also brought into this, and Honeywell as the main control vendor was providing lower level data directly into the databases. The subsea contractor that was part of a separate contract with Woodside also got access to the document archives. The project manager sums up their experience in this way:

> I think the document archives meant the single biggest achievement of the project. Without an application like that, the project probably would not have been completed in the time that it was.

The project is now into its fifth year, and DocuLive is still being used as the authoring tool through the maintenance and operation phase.

6.4.4 Some Problems in the Utilisation of Collaboration Technologies

Kværner today makes extensive use of collaboration technologies for supporting their distributed project operations. However, there are still problems that need to be overcome for making more effective use of the services.

A major problem is related to the training of a temporary workforce in the use of proprietary systems. Kværner typically hires a large number of agency engineers to work on their projects, and these are often unfamiliar with Kværner's proprietary systems:

> If you get in an agency engineer to do some work on mechanical data sheet, he has never seen this application, with screen forms and menus. He usually knows Excel format from many jobs, and that is all he wants to use. So there is quite a mountain to overcome with training and getting the actual engineers to accept it.

Further, this training often has to be conducted at the job rather than at the outset, as the project budgets do not include training. One way of trying to overcome this challenge has been to appoint superusers. In general, the IT department in the O&G division tends to provide a "layer" of superusers for the projects, assisting with the population of data in the repositories, making modifications to reports, etc. For example, at the peak of the Laminaria project there was a team of 8-10 IT people mobilised on the project as superusers. In comparison, the E&C division tends to apply superusers from the disciplines rather than from IT. The arguments for this are that the superusers then are more responsive to the discipline requirements, and that it is easier to charge them to the project this way.

There have also been problems with getting engineers to populate the repositories with data that may not be of direct use to the current stage of the project, but that are vital for later stages. As experienced in the Laminaria project:

> A mechanical engineer who is responsible for the technical side of procuring a piece of equipment often could not see any benefit directly to himself of some of the data that he was asked to put into the database. But it was much better for him to put it in at the time it was available, because the value of that data towards the end and into commissioning was absolutely immense. So we had the problem that the engineers would tend to fill in only the fields of information they needed and required, and when we got towards the end of the project, the fields that were supposed to be filled and ready for the commissioning people, were missing. So we had a *big* exercise to pour through vendor documents and other information, and there was a huge effort to populate the databases right at the end when the theory is really to trickle-feed them through the job.

Again, training is considered the major means for overcoming this problem. But this is being limited by the costs.

> The problem is that it's a very, very competitive industry and you are trying to cut costs at every corner. And at the beginning of a job you don't want to add an extra couple of hundred thousand dollars for training, you really want to get started and get into the job and win the job and get doing it rather than wasting time, so called, training. It's something that is very difficult to do.

However, Kværner is experiencing that the climate is changing in the sense that several of their clients are becoming more receptive to the need for training, and understanding the importance of it. As a result, many projects now start with several of the client's engineers coming to work in Kværner's offices as a project team. This team may often go away for a couple of days and do teambuilding exercises, which helps to engender a cooperative approach within the team.

A third factor limiting the potential benefit from the document management systems is that despite the advanced search features of the electronic archiving system, some people still prefer using hard copies of the documents.

> A lot of the time when they ask somebody for some information, the response they get back is "the document's in DocuLive, go and find it", whereas in a hard copy world, you would expect somebody probably pulled it off the shelf and say "oh, here it is, I got a copy of the latest one".

So the potential benefit of a "paperless project" has not yet been completely realised. According to a project manager this may also have to do with lack of trust in the technology:

> Also, I think people still don't trust the computer, and they like to have a piece of paper, and they like to write notes on it and put it in their file saver. That's a big thing to overcome.

6.5 Future Challenges

With the reorganisation of Kværner the vision of centres of competence has been realised, in the form of regional offices being particularly strong in specific areas, such as metals in the UK office in Stockton, oil and gas in Perth, Australia, etc. Projects in these areas therefore tend to be channelled to these centres. Although KINET and the engineering applications support distributed operations, however, the clients' preferences for co-located operations may still constitute a barrier to this approach. As expressed by the IT manager in Kværner Australia:

> The idea was to have these centres of excellence and use the network to tie them together. The problem has been though that the clients typically don't want to have remote processing, they like to have the project and project task force in their city. For instance, we did a recent job in Brisbane where the client wanted a lot of the design to be done in the Brisbane office, when this was something that traditionally Melbourne is stronger at doing at that particular plant. So, we do have a lot of client resistance to that vision.

The large variety in projects also makes it challenging for Kværner to be able to apply a best practice approach:

> The difficulty is, with the range of clients that we have, from major oil companies who are used to doing billion dollar projects down to some of the small, perhaps mining companies where this is their first project and they don't know what to expect, but nevertheless being a client they want to dictate some of the way that we are doing things. So we can't necessarily always execute a project in a way that we think is the best way to do it. If they want to do it in a particular city or they want to use their procurement people to do it we have to integrate their procurement team into our design team for instance. And so on, there are hundreds of different combinations and every project's different.

Further, the sharing of electronic archives with third parties as part of Kværner's B2B strategy also creates legal issues that need to be resolved. For example, the project manager for the Laminaria project explains how people still extract drawings and information from the project database to assist them in development of work on other projects.

> Who currently owns the documentation that's in there, and who has the right to go in and take a piece out of the P&ID and say, "oh, that separator configuration is exactly what we need on this other work?" I suppose engineers have been doing that forever, but the client might say "I paid for it, it's my drawing".

Kværner's position is that they maintain copyright on the design, but sometimes they are forced to change this. It is therefore important that this is negotiated at the beginning of a project. Some of the bids are also starting to include information portals as an element, with the question then arising of who should control this portal:

> Sometimes the client wants to control it. The difficulty we have is, at the beginning of the job 99 per cent of the information is being generated by us and it doesn't make sense to send all that out to a remote portal, it's better, in our view, to have it local at least to start off with.

Finally, an important challenge for Kværner in the future will be to increase the use of collaboration technologies for capturing project experience and transfer this among the companies. Although the infrastructure is in place, the use of this for experience transfer between the companies is still scarce, and collaboration between the centres of competence is so far limited to sharing data rather than knowledge. Disciplinary boundaries between the E&C and O&G divisions as well as tight project deadlines act as barriers to this. The high number of agency engineers also contributes to make this a pressing but difficult challenge. As these agency engineers are hired based on their knowledge, they have few incentives for putting in extra efforts for "storing" their expertise with their employer.

For all major projects a lessons learned session is held where lead engineers get together and discuss the experiences from the project. Based on this the project manager develops the "project close up report", which is a key document for capturing the lessons learned. In general, there is increasing emphasis on this and such documents have also started to be put out on the intranet. However, this is not a formalised procedure yet, and there are still examples where there have not been any resources left at the end of a project for developing this report. Further, potential liability issues also tend to restrict the use of these services for documenting project "failures", as these documents may actually be declared as material to be used in court in the case of a conflict with a client or a third party.

6.6 Lessons Learned from the Kværner Case

The case of Kværner provides interesting lessons learned both related to the process of establishing a collaborative infrastructure, and to utilising this technology for supporting distributed projects. The implementation of KINET was slower and more resource demanding than expected at the outset. The top management's vision of large benefits through increased collaboration among the Kværner companies was not shared by all companies. Representing a new way of working, this was difficult to relate to for companies without prior experience with distributed collaboration. To gain user acceptance for a new technology there must be some perceived relative advantage from adopting this technology (Rogers, 1995), it needs to be perceived by the users as an improvement to the existing ways of doing things. Several of the companies were already using alternative technologies, and with most of their collaboration taking place with partners in their local markets they did not experience a need for changing the status quo. The local context of the adopting companies was found to influence the implementation project in several ways, including the political concerns in choosing local suppliers or vendors, or the protective regulations imposed by national telecommunications providers. The large heterogeneity in technological platforms added to the complexity in the implementation process, also illustrating the problems with setting standards in a large, decentralised group as Kværner. Much of the delay in the implementation project was due to the problems with developing a network architecture serving the large variation of companies.

To overcome the barriers in the implementation process, the Kværner implementation team had to adapt the presentation and marketing of the technology to the local culture and needs of the different member companies. Sponsoring of adoption costs was also applied in the form of establishing regional hubs for accessing the network. The rapid diffusion and increase in network traffic after a critical mass of users had been reached shows how the problems in the implementation were mostly confined to the early stages. With more explicit backing and pressure exerted from the Kværner top management, it is likely that a critical mass of adopting companies could have been reached at an earlier stage in the project. The new top management in Kværner is being described as more "computer literate" than the previous one, putting more emphasis on IT as a strategic tool.

The changing role of the IT function in Kværner from a large, independent unit to smaller, localised service providers can be seen as an example of a trend taking place in many companies today. Kværner has chosen to outsource both the maintenance of their network infrastructure and the further strategic development of KINET. Whether the strategic concerns will be equally well maintained by a third party remains to be seen.

Kværner has also experienced several challenges related to utilisation of the collaboration technologies, and the level and sophistication of use has not yet reached the goals set for the KINET implementation regarding effective communication and knowledge exchange among the Kværner companies. In general, several of the problems are related to the competitive climate of Kværner's operations, and the need to keep the costs down. Every project involves a constant balancing act between schedule, cost and quality. Limited time and resources for training of a temporary workforce has restricted more effective use of the proprietary systems. The training also needs to focus more on the cooperative aspects of the use of these technologies. Instead of viewing their jobs as narrowly defined, the engineers need to appreciate how the information they are asked to supply to the databases may be appropriate for the project later on, even though it does not assist them directly in time.

Increased training thus emerges as an important solution to many of the current challenges that Kværner faces related to the use of the technologies. This also requires that the clients become receptive to this need, and allow for this to be included in the project budgets. This is a transition that seems to be taking place, although at a gradual pace. The current preference of many clients for co-located project operations also constitutes a barrier to the deployment of these technologies. This is difficult for Kværner to overcome without attitudes changing on the client's side. The increasing integration with clients and third parties in the projects as part of the B2B strategy can be seen to contribute to this transition. As reflected by the IT manager in Kværner Australia:

> I think that one of the things that these collaboration technologies have shown us is that we're redefining the boundaries between engineers and partners and so on, and we're a much flatter organisation now, we don't have these little sort of islands so much, there is much greater integration and job definitions are changing. And these collaboration technologies are something that's driving that.

Through the implementation of KINET and their current use of collaboration technologies, Kværner has come a long way towards realising the vision of effective use of IT for supporting global engineering projects. Increased use of the technology for supporting experience transfer and knowledge exchange between the companies represent additional benefits to be realised. This process is still in an early stage, and will require further transition in work practices for Kværner. Further, as the clients' awareness of the potential benefits from distributed projects increases, the use of collaboration technology will become even more important for Kværner's operations.

Acknowledgements

The author is grateful to the interviewees in Kværner, who kindly shared their valuable experience.

7

Implementation of Data Conferencing in The Boeing Company

Steven E. Poltrock and Gloria Mark

7.1 Overview

Between 1996 and 2000 The Boeing Company implemented a data conferencing service to support collaboration within the company and with external partners, vendors and customers. By the end of 2000 usage of this service exceeded 40,000 user hours per month. Employees use this service both in conference rooms and at their desks to look at and interact with documents together while speaking to one another over the telephone.

Boeing's applied research group, where the first author works, played a significant role in the introduction of this technology and its transfer to the company's information systems organisation for implementation and support. We continue to participate in planning the evolution of this service, and we have studied how its usage grows, ways that people use it, and the impacts of its use on the company.

In this paper we report how we developed and transferred the data conferencing service and some of its impacts. The service was developed as a prototype in an applied research group that investigates emerging technologies and their potential application to business needs. We consider the role of the investigation phase in the successful implementation of this service. When experiences with the prototype demonstrated its potential value for supporting collaboration, we initiated its transfer to the company's information systems organisation. Transforming a prototype into a robust, reliable service intended for use throughout a large multinational company proved far more challenging than we expected. Some of these challenges were technical, but the greatest challenges were organisational. We conclude with the benefits and shortcomings uncovered in studies of its use.

7.2 Data Conferencing

Data conferencing is a computing technology that enables people to communicate and share information with one another synchronously. The International Telecommunications Union (ITU), which also establishes standards for telephone communication, established the T.120 family of standards defining data conferencing functionality and technical protocols (see DataBeam, 1997). Much like using a telephone, a person can use a data conference application to call another person, or they can call a conference bridge to join a session with many other people. Instead of dialling phone numbers, people are called by entering their network addresses, or more commonly they are located in an online directory that records names and network addresses. When two or more people are in a data conference session, they can

- work simultaneously in a shared whiteboard;
- share application windows from one person to all other participants;
- allow other participants to interact with the shared applications;
- enable short text messages to be sent to one or all participants;
- send files to the other participants.

The primary components of the data conference service implemented at The Boeing Company are Microsoft NetMeeting as a Windows-based data conference application, a DataBeam neT.120 data conference bridge, and a Microsoft Internet Locator Service (ILS) as a directory of data conference users. Data Connections Limited developed a compatible data conference application for UNIX called DC Share, and versions of it are available for UNIX computers from Sun Microsystems, Hewlett Packard, Silicon Graphics, and IBM, as well as computers running the Linux operating system. Data conference products are available from other companies also, and now data conference services are available on the Internet.

7.2.1 The Support Environment

An essential element of our successful implementation of data conferencing is a robust support environment provided by Boeing's information systems organisation. People are assigned the jobs of product manager for each of the key technology components, and they are responsible for deploying and supporting these environments. The product managers are our principal channels of communication with the vendors who develop data conference technologies, and they are responsible for evolving the service. They lead evaluations of new products and product versions; they recommend whether to replace an existing product with a new one; and they customise the products to integrate them in the Boeing computing environment.

The product managers develop websites where employees learn about data conferencing and how to use it effectively, they answer questions about these capabilities, and they act as the voice of the company when communicating our requirements to software vendors. The website includes training materials that show step-by-step procedures for calling other people or a conference bridge. The website also includes suggestions about etiquette for online meetings, such as a suggestion that people say their names when they begin speaking so everyone will know who is talking.

The data conferencing environment has been carefully constructed to provide service 24 hours a day, 7 days a week. The servers are monitored to ensure they can meet the growing demand, all data are backed up regularly, and backup servers are available for immediate replacement in case a server ever fails.

7.3 The Investigation Phase

Boeing Phantom Works is the applied research and development organisation of The Boeing Company, and Mathematics and Computing Technology (M&CT) is the organisation within Boeing Phantom Works responsible for investigating emerging computing technology and its potential for improving Boeing's processes and products. The first author began investigating collaboration technologies at Boeing more than a decade ago, exploring the capabilities of new technologies and learning through observations and experiments how these capabilities address business needs (Poltrock, 1996). These investigations included general-purpose groupware products such as Lotus Notes and Domino, document management technologies, workflow management technologies, data conference technologies, videoconferencing technologies, and web-based team collaboration environments. All these collaboration technologies are used in Boeing today, but none has been as widely adopted as data conferencing.

Our long-term investigations of collaboration technologies were critical to the successful implementation of data conferencing by preparing us to act successfully as change agents (Rogers, 1995). We knew what capabilities were needed, how they would be used, what benefits they would offer, and who would most likely be early adopters. We were prepared to recognize when the technology achieved the maturity necessary for implementation, and by rejecting premature technologies we had established our expertise and credibility.

7.3.1 Collaboration Requirements

The Boeing Company develops large integrated systems including commercial aircraft, space and communications systems, and military systems. Large system development requires collaboration among geographically distributed people and organisations. Many of our development programmes involve companies distributed around the world. Some programmes are funded by multiple nations, such as the International Space Station and other space and defence-related programmes. Some programmes develop system components at multinational locations to take advantage of localised skills and cost differences. Of course, large-scale global collaboration is not limited to our company or industry. For example, Herbsleb et al. (2000) reported that people located at multiple sites in four countries distributed across three continents all worked together developing software for Lucent switches.

Necessarily, much of the collaboration in these programmes is accomplished asynchronously without any two people speaking to one another. The work and the collaboration among programme participants are carefully scheduled and managed. The staff is organised in a hierarchy of interrelated teams (see Poltrock and Engelbeck, 1999) that sometimes span locations, and the interfaces between teams are well defined. Contracts structure much of the collaboration between companies, specifying what will be delivered from one company to another and when.

This chapter focuses on synchronous collaboration, which may also be scheduled but is generally much less structured and organised. Despite the difficulty of working together synchronously across different time zones, its importance is increasingly recognised. Rocco et al. (2001) found that programme performance depended strongly on trustful relationships among the staff. These relationships are developed through synchronous communication, especially face-to-face meetings.

Recognising the importance of synchronous collaboration, many Boeing programmes collocate key staff members during the early phases of a programme. While collocated they develop detailed plans for the later work, learn common processes, and develop personal relationships. As the programme unfolds, people often meet again for key events such as major design reviews. These face-to-face events are, however, expensive and too infrequent. Programme teams need frequent synchronous collaboration in order to coordinate their work.

Prior to implementation of the data conferencing service, synchronous alternatives to face-to-face meetings were telephone conference calls and videoconferences. Telecommunication companies provide telephone conference call services universally, and these services are widely used. Unfortunately, voice is not an adequate medium for all communication. Engineers want to discuss details of designs and specifications that are not easily described in words. Videoconferences enable people to see and hear one another, and generally people report liking the experience, but there is little evidence that video aids in the performance of technical work (see Finn et al., 1997). When engineers participated in videoconferences, they often pointed the camera at documents they wanted to discuss and were disappointed to discover that the documents were not readable. In addition, videoconference technology and the required network bandwidth were expensive and therefore not readily available. We should note that advancements in both videoconferencing and network technologies may soon make videoconferences more productive and cost effective.

Data conferences provide a lower cost alternative to videoconferences, and they display exactly the information of greatest importance: the designs, specifications, and any other digital data.

7.3.2 Technology Exploration

In 1989 we began exploring uses of data conference tools for the UNIX operating environment such as XTV (Abdel-Wahab and Feit, 1991) and Shared X (Garfinkel et al., 1989). As research prototypes, those technologies allowed us to demonstrate the capabilities and discuss potential future uses. We anticipated that engineers would want to use data conferencing to collaborate while using computer-aided design (CAD) tools, but we soon learned that they had relatively little interest in that capability and there were some challenging technical issues. We learned that engineers did not want to reveal unfinished CAD models because the interfaces between models are critically important. They worried that an engineer looking at an unfinished model could reach incorrect conclusions about the interfaces. In addition, these early prototypes leveraged features of the X Windows display environment, but our CAD tools used special display hardware that bypassed the X Windows commands; the CAD displays could not be shared. Furthermore, these early prototypes were not reliable or robust enough to place in production.

Early in our investigations, we observed that many Boeing employees wanted to collaborate on documents, both at a distance and in face-to-face meetings. Documents were central to much of the collaboration within product development teams. They developed statements of work, work breakdown structures, schedules, specifications, action item lists, and many other documents that served to coordinate across people and time. We experimented with Aspects, a commercial product that supported collaborative authoring on Macintosh computers (Poltrock, 1996). Two or more people could edit the same document concurrently through Aspects. Unfortunately, the demand for this capability was greatest among the authors of very large, complex, highly structured documents, and Aspects did not adequately support such documents.

As Microsoft Windows grew in popularity, demand increased for collaboration capabilities on the Windows platform. One of the first data conference technologies for Windows was Smart2000, developed (but never released) by SMART Technologies. Smart2000 adopted a replicated data conference architecture (see Greenberg and Roseman, 1999) in which identical software runs on a group of computers, and inputs made on one computer are sent to all computers in the group. Suppose, for example, Microsoft Word is running on all the computers. A user could type on one computer, and the typed input would appear on all computers simultaneously.

The major shortcoming of the replicated architecture is that it requires keeping all the computers synchronised, and this can be difficult. We experimented with Smart2000 in 1995, equipping a conference room with six computers, one for each participant in a face-to-face meeting (Poltrock, 1996). We ensured that all six computers had identical software and data files to minimise synchronisation problems, but we encountered problems nonetheless. For example, during a team review of a document written in Microsoft Word, one person moved the cursor by holding down an arrow key instead of using the mouse. Holding the arrow key moves the cursor a distance that depends on how long the key is depressed. The other computers observed different durations because of small random variations in network latency, and the result was that the cursor arrived at different locations. Subsequent revisions to the document were made at different locations in each document, and the participants in the meeting, who each saw only their own copy of the document, were confused. Once such asynchronies began, they were very difficult to recognise and correct.

Data Connections Limited developed a technology called DC-Share for data conferencing that avoided the synchronisation problems we experienced with Smart2000. They adopted a non-replicated architecture (Greenberg and Roseman, 1999) similar to the architecture of the earlier UNIX prototypes. DC-Share captured the display commands of an application running on one computer, encoded the commands, and sent them to other computers in a collaboration session. Those computers decoded and executed the display commands. Everyone in a session could see the windows of an application that was running on just one of the computers, and they could interact with that application, although only one person could interact at a time. DC-Share was built on the T.120 infrastructure originally developed by DataBeam (1997) and adopted as a standard by the International Telecommunications Union (ITU). The scheme developed by Data Connections Limited for encoding the display commands was later integrated into the T.120 standards as T.128.

Early in its development, we experimented with DC-Share to support two-person discussions about documents. We found pairs of people in Boeing who were collaborating on documents while working many miles apart. They travelled frequently to meet face-to-face and discuss the documents, and they were intrigued by the opportunity to work together without travelling. Using DC-Share, they talked on the phone while looking at the documents together. They experienced many difficulties using this technology, however. Some of those difficulties were associated with its stage of development and with the state of our infrastructure. Configuring the computers to call one another over our internal network proved difficult because the network services encouraged client–server communication, not peer-to-peer communication. For example, there were no directory or name services providing the computer addresses of other users. We knew these difficulties could be overcome by technology improvements. We were more concerned by the difficulties our users had in changing their work habits. They were accustomed to marking up paper and reviewing those revisions while sitting together. Few people knew they could mark up the digital medium, and they preferred working with paper and pen. This new mode of work was unfamiliar, and none of our users embraced it. They tried the technology and then returned to travelling to face-to-face meetings.

7.3.3 Our Prototype Data Conference Service

In 1996 we acquired and began testing early versions of the product later released as Microsoft NetMeeting. Based on the T.120 standards and the T.128 extensions to the standard developed by Data Connections Limited, NetMeeting provided several methods for a group of people to collaborate with one another in a data conference session. As in DC-Share, one person could run any application and share its windows with other participants and pass control of the application to other participants. In addition, they could collaborate synchronously in a shared whiteboard (a replicated architecture application), send chat messages to one another, and send files to one another. Later versions added audio and video communication in accordance with the H.323 family of multimedia communication standards (DataBeam, 1998).

Because of our experiences with users of DC-Share, we were concerned that people would have difficulty finding one another and finding meetings of people over our network. We acquired the DataBeam neT.120 data conference server to give people a place to meet. A data conference server acts as a bridge, tying together all the participants in a virtual meeting. The conference server can be viewed as a collection of virtual conference rooms that can be scheduled and where people gather for a meeting. Meetings are scheduled using a web browser and an http form. Using NetMeeting, people simply call the conference server and enter the name of the meeting they wish to join.

Microsoft soon offered an alternative solution to the problem of finding other users. They developed an Internet Locator Service (ILS) that serves as a dynamic directory of NetMeeting users. When started, NetMeeting registers with an ILS providing the user's name and e-mail address and the computer's network address. If not refreshed, the information is deleted from the ILS. NetMeeting repeatedly refreshes the information until it is terminated. Through the NetMeeting directory, each user can browse or search a list of all other people registered with the same ILS.

Note that our prototype data conference service provided two ways for people to initiate a conference, which has been an ongoing source of confusion. The NetMeeting directory shows them all the people registered with the ILS, and they can easily call any of these individuals. But a person who wants to join a team meeting will not find the team listed in the directory and may not know which person to call. The data conference server provides virtual team meeting rooms, but users must call the conference server (which is not listed in the directory) by name, then find the meeting or enter its name.

7.3.4 Early Experiments Led to Redesigned Server Interface

We quickly discovered that people had difficulties using the web interface for the DataBeam neT.120 data conference server. Researchers at Motorola (Day-Ryan, 1997) shared the results with us of a thorough study of the usability of this conference server. They found that novice users were able to perform many essential tasks correctly less than half the time, and experienced users were also frequently confused. Another company that used this product told us that those people who scheduled meetings on their server were given a full day of training. We anticipated that success at Boeing would require that the technology be usable without training by the same people who schedule our conference rooms, the office administrators. We simplified the interface by removing some of its fields, and we added validity tests for the fields that remained. As described below, we made many more revisions to this interface later when transferring this technology to our information systems organisation. Our modifications made the interface more usable, but equally important, they established our expertise and ability to improve the service.

7.3.5 Growing Use Through Discovery Inside and Outside

We began demonstrating our prototype data conference service in 1996 to people within Boeing known to be interested in these technologies, and a few dozen people used it during the first half of 1997. In May 1997 we invited everyone in our research organisation (about 250 people) to begin using the prototype data conference service and to invite their colleagues in the company to use it with them. We had established a website where they could download NetMeeting 2.0 (which had been released just a few days previously) and learn about both NetMeeting and the conference server. We gave demonstrations to the research staff at technical meetings and provided a little technical assistance to those few researchers who asked for help using it. Our environment was primitive but functional. Both the conference server and ILS were running on old, slow 486 computers, but their performance was adequate for the very small demand placed on them.

Usage of data conferencing increased slowly during the year following our invitation. Early adopters in our research organisation learned about the technology and told their friends about it, but even a year later we found that many of our research colleagues knew nothing about it. We also proselytized the benefits of data conferencing in presentations to information systems organisations throughout the company. Our presentations added to the early adopters. In addition, some people learned about data conferencing from trade magazines or from the Microsoft website. They were surprised and pleased to discover that a prototype infrastructure existed to support their use of this technology.

In July, two months after we launched our prototype, a virtual meeting was held that triggered the next phase of our implementation. The executive support staff acquired NetMeeting from our prototype service and persuaded our chief executive officer to try using it in a meeting held simultaneously in the corporate boardroom and in one of the presidents' conference room. The executives immediately recognised the potential for supporting meetings among members of the geographically distributed executive council. We encouraged them to use the neT.120 data conference server also to increase the reliability and security of the conference, but initially they only used NetMeeting, hosting the session on a computer in one meeting room. A few months later their host computer started a scheduled disk backup in the middle of a meeting, disrupting their data conference. From then on, they used both NetMeeting and the neT.120 conference server for the executive meetings.

7.4 Technology Transition

In 1996 we felt ready to begin diffusing this innovation (see Rogers, 1995). The technology had reached a reasonable level of maturity and it was well priced. From our perspective as change agents, we needed to follow two paths concurrently: (1) Introduce it to early adopters who would establish its usefulness and tell others about it; and (2) transition its support to the central Information Systems (IS) organisation. Early adopters soon learned about the technology from our presentations, our website, and news stories. Adoption could not expand, however, until the products were approved for use within the company and IS began supporting them.

Each major division of The Boeing Company has its own IS organisation that is responsible for developing the division's applications and supporting its programs. In addition, a central IS organisation provides services that are common to all divisions and coordinates with the division IS organisations. The central IS organisation manages the e-mail servers and the intranet firewalls, for example. It leads the development and evolution of the IS architecture. Technology transition requires engaging both the central and the division IS organisations. The central organisation provides the common infrastructure and the division organisations encourage its adoption.

Engaging IS in supporting data conferencing was slow initially, but accelerated rapidly in 1997. IS initiated a Virtual Collocation Project to expand support for collaborative work, and the first author was a member of the project team. This project provided a forum for discussing a wide range of issues including etiquette for virtual meetings, videoconferencing, and support for multimedia e-mail. The project focus narrowed substantially, however, when the IS community learned that the highest-level executives were using our prototype data conference service. When we informed them that the corporate executives were holding business meetings using an old 486 computer located in the hallway in M&CT, the discussion became very focused indeed. The value of executive sponsorship is well known in the collaboration technology research and development community, but we observed that executive use of a technology is almost as valuable as their sponsorship.

We naively expected to transfer support for data conferencing quickly and easily. The prototype service had been running for months with little or no problems. Microsoft was giving away both NetMeeting and the ILS at no cost, and we had already purchased the neT.120 server. We thought the transition would take a matter of weeks, but the service was implemented as a pilot at the end of the year and became a fully supported production service almost a year later.

7.4.1 Production Services Are Complex

As we learned, it is challenging to develop a new production service for a company with more than 200,000 employees. There are an extraordinary number of issues to consider and complex processes to follow. As in any large well-run company, the information systems architecture is carefully managed, and new technologies must be carefully evaluated for their

potential impact on the infrastructure and other applications. NetMeeting, the ILS and the DataBeam neT.120 server each were thoroughly tested.

Network Issues

Network impact was a critical and controversial issue. The IS organisation tested network usage as each of the functions were used in small and large meetings. They found that network usage spiked when people joined a meeting, especially when using the NetMeeting whiteboard, but the bandwidth was not large enough to cause concerns. Network impact was controversial because NetMeeting included audio and videoconferencing over the network, and these functions used much more bandwidth. Once NetMeeting was approved, everyone would be able to acquire and install it and many might use these features, possibly compromising network performance. We pointed out that people would not use NetMeeting audio and video if its performance was poor because poor audio and video are worse than none at all. NetMeeting was eventually approved for use, but its audio and video were not approved.

Voice and video were the greatest concern in terms of network impact, but some groups worried about the impact of other features. These concerns were put to rest by helping people understand that the key factor in network load was the utilisation rate for NetMeeting (number of clients actually in a call), not the number of clients installed. We used an analogy to telephone systems; a company may have several thousand telephones installed, but only a fraction of those devices are off-hook and using system capacity at any given time. Our NetMeeting infrastructure was analogous to a telephone system. NetMeeting only places a load on the network when the client is in a call and sending information updates (e.g., when changing PowerPoint slides). We only needed to be sure that our networks could handle the typical utilisation rate of NetMeeting. This approach proved itself in actual usage. With over 30,000 NetMeeting clients installed, only 200–250 users are actually in a call at any given time. The actual network impact is very small and not a significant factor to the Boeing network.

Teleconference Issues

The telecommunications staff was also concerned about the impact of the data conference service. If people began to use this service instead of driving to face-to-face meetings, then they would use the telephone more and demand more teleconferences. They sought our help anticipating the growth so they could plan for increased teleconference capacity. However, their request for funding to expand teleconference capacity was denied because of doubts that data conferencing and teleconferences would expand as rapidly as predicted. Of course, conference usage did expand rapidly, and providing the teleconference service became a significant challenge for the IS organisation.

Product Managers as Change Agents

Product managers were identified for each of the three products (NetMeeting, ILS, DataBeam neT.120), and they developed support plans for each of them. From our perspective as change agents, the assignment of product managers was a major event in the technology transition because they would take responsibility for some of the work we had been doing. Product managers act as change agents by leading development of production services based on the selected products and evolving the services as new products emerge. They would take responsibility for the technologies that comprised our prototype. They differ from the change agents normally found in diffusion of innovation research (Rogers, 1995) in that they are not allowed to advocate their technologies through advertisements.

The NetMeeting product manager constructed a website where employees could download a custom version of NetMeeting configured to work with our ILS servers. This site also contained information about effective virtual meeting behaviour, information about the neT.120 conference servers, and links to other useful information. The ILS and neT.120 product managers purchased two computers for each service, a primary and a backup in case of a failure. They developed procedures for maintaining the services, accumulating usage data, and backing up the data. They also analysed the scalability of these servers and initiated discussions with Microsoft and DataBeam regarding how their technology would scale to support the thousands of users we expected and have now observed.

Boeing provides a hotline that employees can call for computing support any time. The product managers also developed instructions for the hotline staff. They developed decision trees that help the hotline diagnose problems users encounter and give the users appropriate guidance. Today the hotline staff often uses the data conference service to show employees how to use applications.

Security Issues

Security was the one issue that had a direct impact on us as the developers of the prototype service. None of these products provided the level of security that we wanted to offer anyone discussing business issues. At the time, NetMeeting provided no encryption or authentication. Anyone able to capture the data packets from our network could decode them and see all the information presented in meetings. Furthermore, users configured NetMeeting by entering their names, and no special skills were required to enter someone else's name. Later versions of NetMeeting addressed these problems to some extent.

The neT.120 conference server provided some options that could strengthen security. By default all meetings were publicly visible, but the user who scheduled the meeting could choose to make it hidden. The neT.120 website displayed a list of all public meetings but not the hidden meetings. All participants had to know and enter the name of hidden meetings. In addition, a password could be defined that users had to enter when joining a meeting. These choices involved a tradeoff between usability and security. People could join meetings more easily if they did not have to enter meeting names and passwords. Instead, they could join a meeting by simply clicking on its name. The problem, of course, is that everyone would see and could join any or all of the listed meetings.

We presented these issues and some alternative ways of balancing security and usability to the IS security staff. They were primarily concerned about meetings that could be accessed over the Internet by non-Boeing employees because of the risk that someone could gain access to sensitive information. They ruled that all those meetings should both be hidden and have passwords to provide maximum security. Unfortunately, the conference server provided no way to segregate meetings that could or could not be attended by non-Boeing employees, so we enforced this ruling for all meetings, regardless of the expected attendees. We redesigned and rebuilt the neT.120 server web interface to make all meetings hidden and require passwords for all meetings, with some negative consequences for usability.

7.4.2 Re-designed Web User Interface

To maximise usability we radically simplified the web interface of the neT.120 server resulting in the design shown in Figure 7.1. We repeatedly modified and tested the user interface until office administrators were able to schedule and modify meetings without any training. Our ability to construct a usable interface was strongly constrained by the requirement that we use DataBeam's CGI scripts without modification. We could change the appearance of the web pages but not the data sent to or returned from each form.

Our modifications to the conference server web interface were very successful in that people throughout the enterprise were able to schedule and join meetings with no training and relatively few problems. The iterations in our design reduced the frequency of common problems such as misspelled meeting names and incorrect meeting times. Nonetheless, this user interface and our modifications continued to be a source of

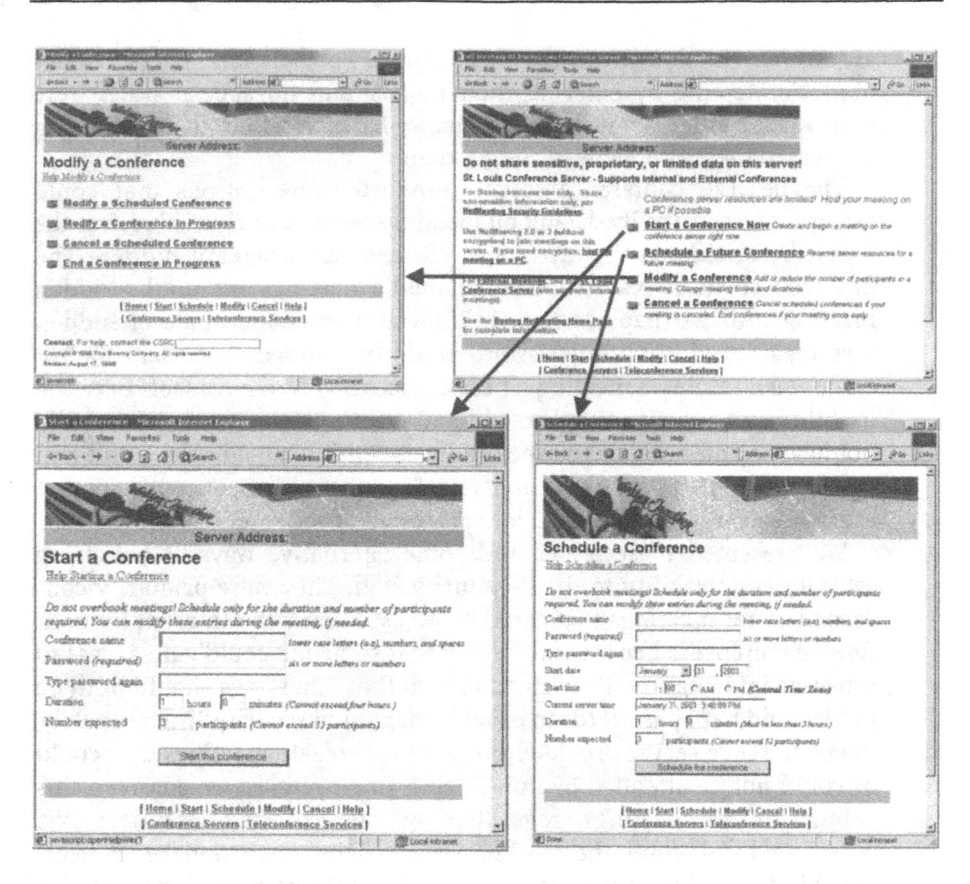

Figure 7.1 Re-designed web user interface for NeT.120 conference server.

problems for several reasons. As noted earlier, the DataBeam server was not well integrated with the Microsoft products. The server itself and meetings on the server were not visible in the ILS, and to join a meeting on the server a user had to enter the name of the server, the name of the meeting, and a password. Many employees never learned how to use the conference server because of its lack of visibility within their NetMeeting interface.

Scheduling meetings on the server was relatively easy, but simple integration with our enterprise e-mail and calendar environment would have made it much easier. Employees scheduled all other meetings, including conference rooms, with Microsoft Schedule+ or Microsoft Outlook, but to schedule a data conference they had to access the conference server's web pages. Furthermore, the conference server would not support repeated, regularly scheduled meetings, so they had to remember to schedule the meeting over and over again. And when they finished scheduling the meeting, they had to manually enter all the relevant information in an e-mail message sent to all the meeting attendees. The conference server did not generate a summary of the meeting data that could be copied and pasted into e-mail messages.

Many of these problems could have been corrected by constructing a new web interface to the neT.120 conference server, but our contract/agreement with DataBeam did not allow us to revise their CGI scripts. We could only change the local behaviour and appearance of the interface through html and javascript changes. We were surprised to learn that even these changes could become a source of future problems. When DataBeam released upgrades of the conference server, they might include an entirely new set of web pages and CGI scripts. We would have to decide anew how much we could use and what modifications, if any, were required.

7.5 Organisational Issues

Adoption of data conferencing within The Boeing Company was influenced by many organisational factors. We have already observed that long-distance multi-organisation collaboration was the principal business requirement driving its adoption. In addition, adoption of our data conference prototype by the executive support group triggered and accelerated transfer of the technology to the IS organisation. Other organisational factors played major roles in its adoption, in some cases accelerating and in other cases slowing adoption.

7.5.1 Rapidly Growing Demand for Virtual Meetings

Mergers with Rockwell North American and McDonnell Douglas in 1996 and 1997 led to rapid growth in collaboration among people too far apart for face-to-face meetings. These mergers approximately doubled the number of employees and radically changed their geographic distribution. Previously, about 80 per cent were distributed around the greater Seattle area. Most employees could attend meetings at any other site by driving less than 90 minutes. After the merger, only about 40 per cent of approximately 235,000 employees worked and lived in the greater Seattle area; others were distributed in 27 states, with large concentrations in California, Missouri, Pennsylvania, Arizona and Kansas.

The merger minimally affected many projects and programmes, which continued working with existing staff, suppliers and partners. Some teams and organisations, however, quickly needed to find ways of working with people at other locations; in some cases with people they had never met. For example, prior to the merger, executive meetings were held face-to-face at corporate headquarters, but following the merger the travel and time costs of collocated meetings were prohibitive. The executives' support staff was searching for technologies that would help to address this problem when they found our data conferencing prototype.

Many distributed teams were created in 1997 to define how the newly merged company would operate and how enterprise services would integrate. Existing enterprise-wide teams were expanded to include members from other locations. Many teams struggled with these expansions, and some teams considered terminating their regularly scheduled meetings because of the difficulty of accommodating remote participants. These teams were relieved to discover that technology existed that could help them all see the same information while talking to one another on the telephone. The pain they were experiencing made them willing to invest the time required to learn this technology and the money required to equip their conference rooms with computers, projectors and speaker phones. The timing of our prototype was fortunate.

7.5.2 Conflicting Objectives Complicated the Service Design

Redesigning the conference server user interface was an exercise in balancing competing objectives of the users, the IS community and the vendors. As change agents, we sought to understand and accommodate all their objectives. While redesigning the service we observed that tension between these communities would be a continuous challenge (see Engelbeck and Poltrock, 1998). Consider the central objectives of each community and how they were at odds with one another.

Our users expected to be able to work easily with people at other sites. They expected the service to be reliable, secure, easy to use, require little or no training, appear professional and be responsive. Of course the end users cannot directly participate in planning a new service intended for use by everyone in the company. Interests of the user community were represented in the Virtual Collocation project by division IS organisations and coordinated by a service manager, and the objectives of the end users are not always identical to those of their representatives. A service agreement was written that defined service availability and requirements. This agreement specified the expected performance of the service in terms of properties such as security and reliability that are readily quantified and very important to both the IS community and the users. It did not, however, specify properties such as usability and appearance that are harder to quantify but may profoundly affect users' ability to use the service effectively.

The IS community sought to provide a reliable, secure, stable support environment at low cost. During a discussion of design options, one analyst remarked, "I vote for the option that requires the least support", a common view within the support organisation. They were driven by their customers to keep their costs as low as possible. This objective was at odds with the interests of some end users who wanted more expensive capabilities. For example, some people wanted to use the audio and video features of NetMeeting. The IS community was concerned that video would significantly increase network traffic, requiring expensive network upgrades, and that cameras would have a high installation and maintenance cost. They decided not to offer video as a part of the service and to discourage its use until it becomes much cheaper.

Vendors necessarily compete to advance their technologies and differentiate their products. Vendors placed more emphasis on enhancing their own products than integration with other products. For example, NetMeeting provided an ILS directory service but no integration with conference servers that would help users find scheduled meetings. Innovative functionality often impeded our ability to integrate products. We had to modify the user interface to the conference server to hide capabilities from users that were not supported in NetMeeting. These vendors also employed different underlying semantics. A valid data entry in one application could be invalid in another. For example, conference names were case-sensitive in neT.120 and case-insensitive in NetMeeting.

The IS community integrates and packages vendor products into a service. Often this service package will have its own internal name but that name may never be visible to users who see only the names and terms provided by vendors. Vendors use unique icons, logos and terms to give their products a unique identity. The help pages of the neT.120 server, for example, used terms and referred to other DataBeam products that did not exist in our service. Their user interface and their help pages directed users to call DataBeam for support. These promotional devices can compromise the usability of a service: terms are inconsistent, icons are inconsistent and advertised functions do not work.

As an internal service provider, the IS community acts as the vendors' customer. The IS community seeks to maintain a stable low-cost service, while vendors seek to sell more products by adding capabilities to each new release. The IS community must evaluate new releases in terms of their impact on the entire service. New releases can require substantial service changes and require costly integration efforts that may outweigh the benefits of new capabilities. We repeatedly evaluated each new release of NetMeeting and the neT.120 server to determine whether the benefits of the new release were worth the cost of integrating the products, developing new training material and distributing it to all users.

7.5.3 Organisational Obstacles to Transitioning New Technologies

In 1996 when we began developing our data conference prototype we also began planning its transition to the information systems organisation. We were confident that a demand existed for this technology, and we wanted to begin preparing for its implementation throughout the company. The means to accomplish this was surprisingly elusive.

The information systems organisation had product managers and support structures for every software product and service that it supported, and they were charged with maintaining and evolving these products and services. Had we come to them to propose an improvement to any of the existing products or services, we would have found a ready audience and established processes for working together. Our data conference service, however, was not closely related to any of the existing products and services. Conceptually, data conferencing was closest to the existing videoconferencing services, but those services were organisationally distant. The videoconferencing services were managed by a part of the organisation responsible for telecommunication technologies, not computing technologies.

Data conferencing was a new service, and there were organisational obstacles to new services, especially one that arose within a technology organisation. The IS organisation responded to its customers' demands for services and their willingness to pay for these services. These customers generally demanded improvements in existing services and reductions in costs. They often demanded new capabilities closely related to their core capabilities, such as improvements to computer-aided design systems. They never demanded new capabilities that they did not know existed.

Transitioning the technology to the IS organisation required educating them and their customers and developing the demand for these emerging capabilities. We demonstrated the technology to people in each major division of the company, encouraged them to use it, helped them to demonstrate it to their management, and encouraged them to ask for this new service. The intense need for this technology was an immense help in overcoming this challenge, but still the process seemed slow. The service could not be transferred without funding, and funding decisions were made only once each year.

In 1997 an IS manager formed the Virtual Collocation project to expand support for collaboration, and it provided a forum for us to inform the IS community about data conferencing and related technologies. This project team included experts in every facet of information systems required to develop this new service, and it acted swiftly to plan and execute its deployment. Until that project was established, we lacked any clear technology transition path.

7.5.4 Organisational Obstacles to Collaboration Technologies

Developing a business case for data conferencing was difficult, and we would expect to face these same difficulties for any general-purpose

collaboration technology. Any new service must offer financial benefits that significantly exceed the costs of creating and maintaining it. Financial benefits are the consequences of increased product sales or reduced product costs. It is easy to construct a hypothetical relationship between collaboration and both sales and costs. Collaboration with customers could increase sales, and collaboration among developers could decrease costs. Quantifying these relationships proved more difficult.

Our colleagues in the major divisions attempted to justify the investment in a data conference service by documenting reductions in travel costs. We had already observed that teams using our prototype service had dramatically reduced the time spent driving to meetings. We noted that many people were participating in meetings that would have required travelling by aeroplane and days away from their offices.

We were surprised to learn that these reductions in travel expenses did not constitute a financial benefit. Financial analysts explained to us that these employees were paid whether or not they travelled, so reallocating their time to more productive activities saved no money. As technology developers we did not understand this argument, though it was clearly based on sound financial principles. Because most collaboration technologies help people work together more effectively across distances, we expect to encounter this problem when we introduce other technologies.

A related problem was the difficulty of finding specific programmes or organisations that would benefit from using the technology and could pay for its development. The company could experience some savings in the cost of travel itself, but the funds used for travel were allocated to each individual manager in the company. If all travel costs had come from a single pool, some of the money in that pool could have been diverted to pay for this service. Everyone wanted data conferencing, but no organisation wanted to pay for the creation of a new service. After long discussions of its costs and benefits, executive management of the IS organisation decided to provide modest funding necessary to initiate the service.

7.6 Benefits and Shortcomings of Data Conferencing

We were motivated to assess the benefits of data conferencing by the difficulties we experienced justifying the costs of implementing the service. We have used several approaches to gather data about the benefits. During the summer of 1998 we observed four teams that used data conferencing in their regularly scheduled meetings and reported these results in Mark et al. (1999). One shortcoming of this observational approach is that benefits and problems are not readily attributed to their use of the technology. We augmented this approach with an experiment in which a team that regularly met face-to-face held some of their meetings using data conferencing instead. More recently, we surveyed users of the data conference servers through questionnaires and interviews about their experiences with the technology (Mark and Poltrock, 2001). These interviews provided anecdotal evidence of the benefits, and we received much more anecdotal evidence by e-mail.

7.6.1 Data Conference Usage

Boeing employees use data conferencing in many ways. One of the most common uses is to give presentations to a geographically distributed audience. The presenter may be in a conference room with a collocated audience and a speakerphone or may be seated alone at a computer with a telephone. In either case, the presenter shares Microsoft PowerPoint so that the audience can see but not interact with the presentation. This usage is so common that some people have expressed surprise when they learn that NetMeeting is not dependent on PowerPoint and can be used to share other applications and collaborate in other ways.

Another common use is to support scheduled team meetings of geographically distributed people. The agenda of these meetings often includes a review of action items and presentations by one or more people, and data conferencing supports these activities very well. Often the team leader runs the application used to track action items and shares its window with the team. They all see the action item list and any updates made to it. Similarly, people generally give presentations by running and sharing PowerPoint. Everyone sees the same slide at the same time, neglecting the brief delays caused by network latencies. Rarely, people *chat* with one another during the meeting via short text messages. They might chat about plans to address a specific action item, for example.

The steps involved in scheduling and joining a data conference are currently somewhat complicated because of poor integration among the components and with our e-mail and online calendar infrastructure. One team member schedules a meeting on the conference bridge using the web interface shown in Figure 7.1, much as someone might schedule a conference room for a face-to-face meeting. The team member specifies a name, password, start time, duration and number of expected participants. In addition, the team member schedules a telephone conference call through Boeing's local telecommunication service. Finally, the team member writes and sends an e-mail message containing the information about both the telephone and data conference to the team. At the scheduled time of the meeting, each team member starts NetMeeting and enters the meeting by calling the conference bridge and entering the meeting name and password. They also call the telephone conference bridge and enter a separate pass code for that meeting.

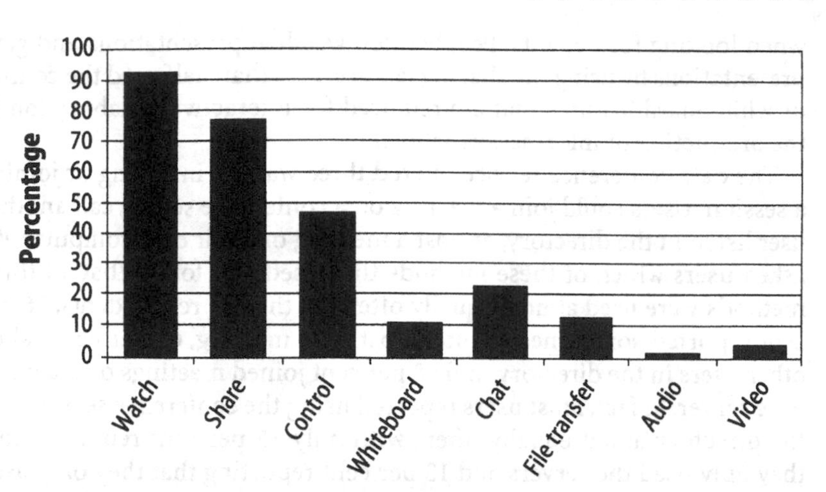

Figure 7.2 Percentage of users who reported sometimes or often using the specific capabilities of the data conference service and its NetMeeting client.

Many people use data conferences for ad hoc, unscheduled collaboration. For example, anyone calling Boeing's internal computing support staff for help using an application may be invited to participate in a data conference. The support staff can see the state of the application, see the user's inputs, and demonstrate proper use of the application. In these situations, they do not use the data conference bridge. Instead, they both start their data conference application (typically NetMeeting), and one person finds the other in the online directory (the ILS) and initiates a person-to-person call. Then they can share the application of interest, and the NetMeeting collaboration feature enables both people to manipulate an application running on just one of the two computers.

We asked 193 data conference users to rate the frequency with which they used the functionality of the data conference service. Figure 7.2 shows the percentage of users who responded that they sometimes or often used a capability of the service. A few users reported that they rarely used any features of the service. All the remaining users reported sometimes or often watching a presentation given by other people, and a majority sometimes or often shared an application running on their own computer so that others could watch. Less than half participated in sessions in which people exchanged control, allowing more than one person to interact with an application. Surprisingly few people used the whiteboard, which is the only way for people to interact simultaneously with the same data. Indeed, 58 per cent reported never using the whiteboard. Similarly, few people used the chat and file transfer features of the service. Audio and video were not formally part of the service, but 4 per cent of users reported sometimes or often using the available video features. These data suggest that we should focus on participation in presentations when looking for benefits. People often watched presentations and gave presentations by using the sharing feature. Less than half used the control or whiteboard features that are required for interactive collaboration in the production of information artifacts.

The data conference service offered three ways of initiating or joining a session. Users could join a meeting on a conference server, call another user listed in the directory, or host a meeting on their own computer. We asked users which of these methods they used and found that all three methods were used almost equally often. Of the 193 respondents, 56 per cent reported sometimes or often hosting a meeting, 60 per cent called other users in the directory, and 68 per cent joined meetings on a conference server. In fact, most users reported using the conference servers and the directory about equally often, with only 15 per cent reporting that they only used the servers and 13 per cent reporting that they only used the directory. An advantage of using the conference server was that people outside the Boeing network could participate in those meetings, and nearly 30 per cent of our respondents had participated in data conference sessions with people outside the company.

7.6.2 Growth of Data Conference Use

Usage data has been collected for our two conference bridges since early 1998. These servers record each time a user joins a meeting, but they do not record the users' identities. Figure 7.3 shows the number of users who

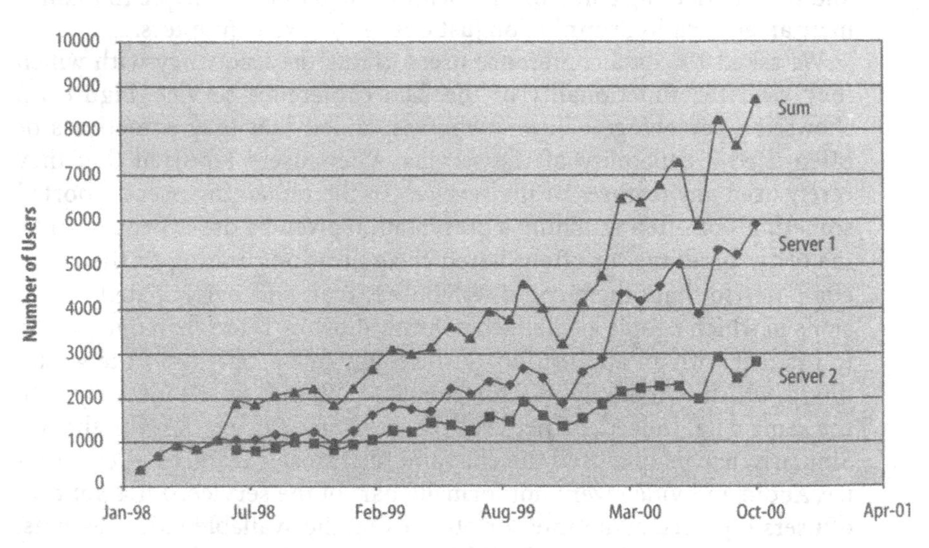

Figure 7.3 The number of users of our data conference servers per month.

joined meetings on each server and the sum across servers, which had grown to 8700 users per month in October 2000. Most meetings hosted on these servers have 3–6 attendees (about 950 meetings per month), and about 350 meetings per month include 7–14 people. Because many people use the service repeatedly each month, the number of people using the service is undoubtedly much smaller than the number of recorded users, but Figure 7.3 clearly shows strong growth in usage of the service.

These two conference servers handle only a subset of all data conferences. Many people call one another directly, which is optimal for two-person data conferences. In addition, one person can host a meeting and many people can call that person. In this case, the host's computer is performing the tasks of a conference server. A more accurate estimate of the total usage of data conferencing can be obtained from the ILS, which shows all the users currently in a data conference. An estimate of the total usage can be estimated by computing the ratio of ILS users and conference server users. The estimated total usage by month is shown in Figure 7.4. This calculation yields estimates of 42,431 user hours of data conferencing in October and 600,109 total user hours of data conferencing during the 33 months between February 1998 and October 2000. Technology diffusion functions are typically S-shaped (see Rogers, 1995), and growth of data conferencing has not yet begun to decelerate.

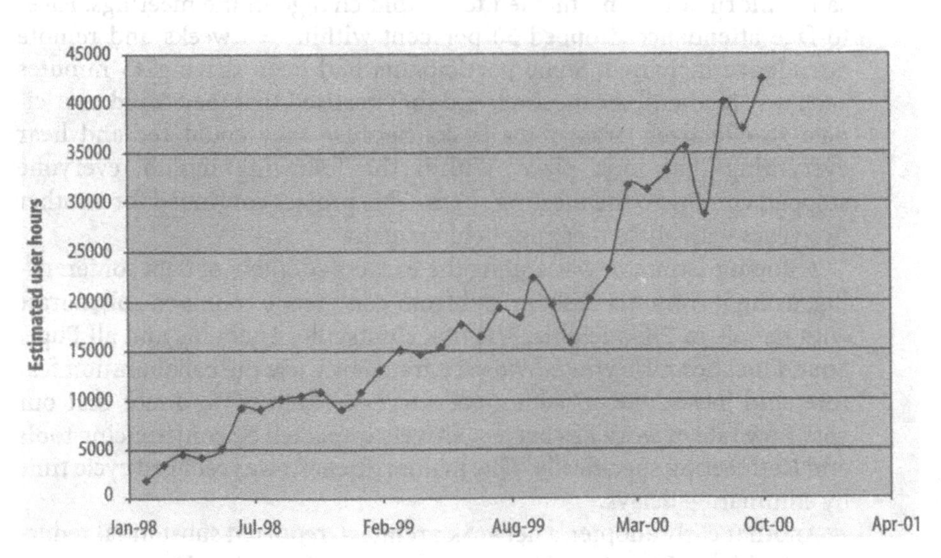

Figure 7.4 Estimated user hours per month of data conferencing including meetings on the data conference servers and calls made through the ILS.

7.6.3 Reported Benefits of Data Conferencing

The data conference service enabled many teams to work together without travelling to face-to-face meetings. An early and interesting example was the Virtual Collocation Project team that was responsible for defining and planning the production service. About 30 people attended the face-to-face meeting in the spring of 1997 that launched this project. Many attended this large meeting to learn what it was about and determine its relevance and importance to their on-going work. Attendance declined steeply over the next two months, reaching a core team of about a dozen people. Soon, however, the team was asked to expand again by integrating participants in Kansas and Missouri. To facilitate their participation, the team acquired a speakerphone for the conference room and scheduled telephone conference calls. The project leader e-mailed or faxed action item lists, presentations and other documents to the remote participants before and after each meeting. We urged the team to use data conference technology, which first required that we install a network drop in the conference room. Later, we discovered that few conference rooms in the company had live network drops, and their installation was an unexpected cost. With the network connection in place, one member of the team brought a laptop to the meeting and used NetMeeting to show remote participants the documents shown in the room. Those of us in the room, however, were not looking at the computer display; we were still looking at transparencies displayed with an overhead projector, and we watched the team leader markup these transparencies with a pen. Next we ordered a projector for the computer so we would all see the same information, and that led to a rapid change in the meetings. Face-to-face attendance dropped 30 per cent within two weeks and remote attendance increased. Some participants had been driving 45 minutes each way to attend the meeting, and they realised that they could participate as effectively from their desks because they could see and hear everything that took place. Within the following month, everyone stopped coming to the meeting room. This project continued for another two years with all its meetings held virtually.

A Boeing estimator was among the earliest adopters of data conferencing, using it from his desktop and from conference rooms to collaborate with people in Philadelphia, Wichita, Huntsville, Anaheim, and all Puget Sound locations. He wrote, "We very frequently use the collaboration feature, and have done so with great success. There is no doubt that our cost/flow rate of work has been positively impacted by conferencing tools and NetMeeting specifically." The primary benefit was reduced cycle time by eliminating delays.

Another early adopter, a network architect, reported substantial reductions in his business travel because of data conferencing. He wrote:

> At the close of business tomorrow I will have gone ONE MONTH without running up an inch of local mileage. I also have not been to Bellevue since July 30. At the height of my old "road warrior" days I would put as much as 300 miles per month on my local mileage form. I didn't plan this change. It just sort of happened serendipitously. I actively participate in Network Synergy Team(s), Network Architecture Board, Wireless TWG, Heritage-ISDS VC Team, Companywide PCS Team, NMD-LSI Team, IDS-WAN Team, SPAC and its architecture sub-teams, etc., etc., but the big change is that *they all facilitate their meetings with NetMeeting and teleconference bridges.* [emphasis in the original]

His experience confirmed the substantial savings that can be achieved by using data conferencing instead of travelling.

The Machining Process Dynamics Team sought to increase the robustness of machining processes by increasing the understanding of the process physics through measurement, modelling, simulation and analysis. This team included hundreds of people across the company concerned with machining processes. The team expanded rapidly after the merger with McDonnell Douglas, greatly straining their ability to have productive meetings. Prior to the merger they had held weekly face-to-face meetings in Bellevue, Washington. They attempted to hold meetings using teleconferences accompanied by information distributed prior to the meetings, but data inconsistency problems plagued the team. These problems were so severe that the team leader considered disbanding the team. Instead, they began using data conferencing and were able to have successful meetings with more people at more locations. Furthermore, people at distant locations in the Puget Sound area began attending some of the meetings through data conferencing, permitting more frequent attendance with lower travel costs.

Phantom Works Mathematics & Computing Technology (M&CT) held two technical reviews every Wednesday morning. Employees at every Boeing location could attend these reviews through data conferencing and a teleconference. Some people from the Puget Sound area participated through data conferencing to avoid local travel costs. People at other sites would rarely incur the expense of travel to these meetings; they simply would not have learned about the work done in M&CT. Their use of data conferencing increased their knowledge of technical advances that could benefit their organisations.

The Collaborative Computing Review Team (CCRT) was chartered in September 1998 by the Computing Architecture Council to recommend architecture, software and hardware products that support collaborative computing for workgroups. The CCRT included 25 people from each major division of the company. Most team members were located in the Puget Sound area, but some were in Philadelphia and Southern California. This team and its sub-teams worked together for more than a year analysing requirements, developing business cases, writing a Request for

Information, analysing vendors' responses, and writing architecture documents. Much of the work was accomplished in meetings using data conferencing to review and revise documents. Most meetings were held in Renton, Washington, where a small group actually met in a conference room. Other team members in the Puget Sound area avoided local travel time and costs by participating through data conferencing. Participation of team members at sites in other states would have been significantly impaired without data conferencing. Data conferencing improved the quality of the CCRT product by enabling participation of team members in other geographic areas and reduced cycle time by enabling the team to accomplish the work in the meetings.

7.6.4 Virtual Meeting Facilitation: Needs and Emerging Practice

We attended and observed the meetings of four teams that adopted data conferencing (Mark et al., 1999). Two of these teams were partially collocated; a group met in a conference room and used data conferencing and a speakerphone to include people at other locations. All members of the other two teams participated from their desks. Many of the meetings we observed began 10–15 minutes late because of the difficulties people experienced joining both the teleconference and the data conference. These problems were repeated over and over again at each meeting. Participants could not find or remember the name of a meeting, its password, the teleconference phone number, or its passcode. Many users were unfamiliar with the technology and could not remember the sequence of steps required to join a meeting or watch a presentation. Our questionnaire data (Mark and Poltrock, 2001) suggests that these problems persist today. Many respondents reported difficulty using data conference technology and long delays at the beginning of some meetings.

Holding effective meetings is difficult even when everyone is collocated, and the difficulty increases substantially when meeting participants are geographically distributed. The leader of a meeting cannot see what attendees are doing; whether they are engaged in the meeting, reading their e-mail, or gone to get coffee. Participants in geographically distributed meetings often engage in activities such as these that they would never do in a collocated meeting.

People have learned meeting behaviours over a lifetime spent interacting face-to-face, and some of these behaviours are ineffective in virtual meetings. In discussions a person who wants to speak unconsciously uses body language to find a good moment to begin. Raising a hand is a simple form of that body language learned in early education. We may not know who is speaking unless we recognise the speaker's voice. Lacking these cues, coordinated turn-taking is much more difficult.

We observed a Scientific Team that developed practices that addressed these problems and made their meetings much more effective. This was a partially collocated team, with one group meeting in a conference room and other people participating either from other conference rooms or from their desks. In addition to the team leader, two new roles emerged in this team that we call technology facilitator and virtual meeting facilitator. The technology facilitator scheduled the data conference session, sent information about it to other participants, helped other people who had difficulty joining the session, shared the application used in the meeting, interacted with that application, and solved any technical problems that emerged. The technology facilitator ensured that little time was wasted because of lack of familiarity with the technology. He summarised his role as follows:

> Every node [site] needs a technical facilitator. The main goal is he must make sure how the details are implemented. The medium must be as transparent as possible. This minimizes the effect on the meeting. To be successful, everyone must be on the same page. The goal is that when the meeting starts, the same page should appear. I try to be in synch with the presenter. I try to be on the same mental page, the view that the presenter wants to give to the virtual group. For example, if the presenter wants to zoom in, I hear the word zoom, and zoom in.

The virtual meeting facilitator sought to integrate everyone into the meeting. He addressed many of the interaction problems observed in other groups. First, he established who was present at each site. Roll call protocols were very informal in the other groups. As with face-to-face group facilitation, he kept order by introducing agenda items and by beginning and adjourning meetings. He continually confirmed that remote sites could see the display, and addressed uncertainty of attendance by checking with remote sites orally.

The virtual meeting facilitator coordinated speaking turns by recognising body language in the conference room (e.g. when someone sat forward or raised a hand) or by hearing an utterance online. He tried to involve remote members in the discussion by asking specific individuals to make sure their questions were answered, referring to people when he believed that they were interested in a topic, or calling on specific people who might have the expertise to answer a question. He encouraged questions at certain points in the presentation, explained to remote participants what was happening at the main site, such as when silences occurred, and kept order during discussions by calling on people. He also ensured that all speakers were identified, especially at remote sites.

Unfortunately, other teams have not adopted this innovation of establishing a technology facilitator and a virtual meeting facilitator. Today we continue to hear complaints about the time wasted explaining how to join meetings and the ineffectiveness of meeting practices.

7.7 Conclusion

The data conference service has become an integral part of business practices at The Boeing Company, and its usage continues to grow. As employees gain experience with it, they think of other ways to use it. Development programmes working with other companies would like to use it to collaborate in the development of complex systems, which will require an infrastructure for secure inter-company collaboration. As this infrastructure evolves, data conferences may become as easy and common as telephone calls are today.

Implementing this technology proved much harder and took much longer than we had anticipated. Some of the challenges we encountered were technical, and generally we knew how to solve those. The more difficult challenges were organisational because we often did not know how to resolve organisational challenges. We conclude by summarising these challenges and how we addressed them.

7.7.1 Technical Challenges and Lessons Learned

- *Assessing technology suitability and readiness.* Through years of experience testing and demonstrating these technologies we had learned about its maturity and the business drivers. We recognised that data conference technology had finally reached a level of maturity sufficient for implementation. We knew people who needed this technology and would request development of this service.
- *Implementation complexity.* Implementing the service was vastly more complicated technically than we had anticipated. It involved assessing network impact, planning telecommunications growth, analysing security requirements and approaches, developing training and hotline materials, creating web pages, managing installations, and addressing reliability and scalability. An interdisciplinary technical team worked together to address these issues.
- *Poor usability.* Scheduling meetings was difficult through the neT.120 user interface provided by the vendor. We radically simplified the interface, restricted input values, added data validity tests, and repeatedly tested the interface with our intended user community.
- *Technology integration.* Integrating products from different vendors is problematic. Each vendor seeks a competitive advantage by offering non-standard features that rarely interoperate. In an integrated environment those features are often lost. When each vendor releases a new product version, the code that integrates the products may need revision. We had no solution for this problem.

7.7.2 Organisational Challenges and Lessons Learned

- *Executive use is almost as powerful as executive sponsorship.* Many industry analysts have spoken of the importance of executive sponsorship for collaboration technologies. We discovered that executive use was almost as effective. Executive use of our prototype service was a key driver of our engagement with the IS organisation to transition the technology to a production support environment.
- *Barriers to new collaboration technologies.* Information systems organisations are structured to support the services that they offer, and adding a completely new technology may be difficult. They cannot provide a service unless their customers demand it, and their customers have little knowledge of how collaboration technologies can benefit them. We demonstrated the technology to their customers and invited them to use our prototype service.
- *Justifying collaboration technologies.* New services require a business case, and developing one for collaboration technologies is not easy. A business case is easy to develop when a technology directly increases product sales or decreases development or manufacturing costs. Working with people in our major divisions we were able to develop a compelling story about likely benefits but could not identify specific, accessible budgets that would profit from the technology.
- *Conflicting organisational objectives.* As designers of the service we found ourselves constrained by the competing objectives of the service users, the information systems organisation, and the vendors. Users want a highly usable, well-integrated service, the information systems organisation and its customers need to minimise the deployment and maintenance cost, and the vendors want to establish competitive advantage and maximise their profit. The resulting service was a compromise among these forces.
- *A need must develop.* Rogers (1995) decomposed the innovation-decision process into the knowledge stage, persuasion stage, decision stage, implementation stage and confirmation stage. We informed early adopters about this technology, kept them informed about its evolution, and persuaded them to try it when it was ready. Passing through these stages takes time, and people move through them at different paces depending on their tolerance for risk. Successful implementation requires patience, repeated communication, demonstrations and direct experience with the technology.
- *Timing is everything.* The rapid diffusion of data conference technology at The Boeing Company was partly, maybe largely, due to the coincidence of mergers that expanded the company geographically. Data conferences could have been adopted as a substitute for local travel,

but the pain and cost of local travel was not great enough to push such rapid diffusion. People adopted data conference technology so they could collaborate more effectively with people far away, and then discovered its convenience for collaborating with people across town.

Acknowledgements

Many people contributed to this implementation and our studies of it. We especially acknowledge George Engelbeck who was a key developer of the data conference prototype. Others who deserve our thanks include David Selkowitz, Vicky Drake, Patia Woods, Kathleen Kelley, Mike McClure, Lisa Pratt, Scott Veasey, Stephen Knapp, David Mueller, Ron Amptmann, Jim Walker, Karen Harlow and Sharon Alford.

8

Discretionary Adoption of Group Support Software: Lessons from Calendar Applications

Leysia Palen and Jonathan Grudin

8.1 Introduction

Although the World Wide Web, Internet and organisational intranets have made computer-mediated collaboration possible for many people, adoption of collaboration technologies in business environments still presents challenges and is often slower or less widespread than anticipated. Technologies focused on supporting groups fall between strictly single-user applications and enterprise systems. Single-user applications are designed with a "discretionary use" model. In contrast, for large enterprise systems, upper management support is considered crucial for smooth deployment and adoption. Which one applies to technologies that support group work? The relatively low cost of an application such as a shared calendar lowers its visibility in an organisation, reducing management attention to it. However, some argue that the complex social dynamics surrounding such technologies still necessitate a managerial mandate for use to occur – the large system approach. Interview studies of electronic calendar adoption in two large organisations found successful, near-universal use achieved without managerial mandate. Versatile functionality and ease of use, associated with discretionary products, were factors leading to individual adoption. Other factors leading to "bottom-up" adoption included the presence of an organisation-wide infrastructure, integration with e-mail, and substantial peer pressure that developed over time.

8.1.1 Technology Adoption by Individuals, Informal Groups, Workgroups and Enterprises

Many opportunities for computer-mediated collaboration are available today on the Internet. People participate in online chat rooms, games and auctions, and they do so at their discretion. As recently as a few years ago, however, the only opportunities most people had for using collaborative information technology to support group activity were at their places of work. There, many found adoption and use of "groupware" to be problematic, and failed efforts abounded. A key question arose: Could a group support application succeed if employees were free to choose whether or not to use it ("discretionary choice") or would it be necessary for management to insist on its use to obtain a collective benefit ("managerial mandate")? This issue arose as large mainframe systems came to be complemented by applications supporting individuals and groups.

An organisation's technology acquisition decisions had been the responsibility of upper-level management, perhaps guided by technology support personnel. Large enterprise systems were expensive and designed to support key business functions. Large systems affected many people, even when the technology didn't suit some individual work practices particularly well. Since local culture and office politics can discourage use in the absence of a mandate to use a system, the expense, visibility and known difficulties faced by enterprise systems led to strong recommendations for a managerial mandate (which still did not guarantee success).

In contrast, similar to single-user applications, many collaborative applications are relatively inexpensive and easy to acquire. They are not tailored to specific business practices. For these, the path to widespread use is not always obvious. Will they follow the discretionary choice model that often works for single-user applications and some Internet-based collaborative applications? Or, as some have argued, do they require a managerial mandate because of the complexities surrounding the social relationships they are intended to support?

Some of these complexities are discussed in Grudin (1994b). Often, group support technologies require additional work from all members but selectively benefit only some, and the collective benefit to the group is not fully appreciated by those who do not benefit personally. In addition, no one person may have the understanding to anticipate impact on the work practices of everyone involved.

Even when everyone will benefit once there is a "critical mass" of users, reaching that tipping point may be difficult. Markus and Connolly (1990) examined social factors in the adoption of "groupware tools" and concluded that mandated use or other top-down measures appeared necessary. In a widely cited case of one such collaboration technology adoption (Orlikowski, 1992), a high-level mandate to use Lotus Notes received a mixed response, but use did continue.

Conflicting signs are present, however. An interesting aspect of Orlikowski's study is that the technical support staff were under no mandate to use Notes but did so before the consultants who were its intended users. Markus (1995) studied mandated use of electronic mail in one organisation and found that *how* e-mail was used was highly discretionary and varied. And e-mail use has often spread without a high-level mandate, although it is risky to generalise from e-mail to other collaboration technologies.

By examining factors that underlie cases of successful adoption of a group support technology, our study addresses the tension between the recognised importance of high-level support for large systems and the lower visibility that collaboration technologies generally receive.

8.1.2 Defining Deployment, Adoption, Success, Failure and Collaboration Technology

Understanding of this topic is hindered by the lack of agreement on the definitions of key terms. The disciplines of human–computer interaction and information systems in particular share several terms but use them very differently (Grudin, 1993).

Writers with an organisational focus use "adopt" to mean "acquire" or "decide to use", whereas writers with a user-centred focus are more likely to use "adopt" to mean "begin to use". To the former it would be possible to say an application was adopted even when no one used it, to the latter adoption implies use. In this chapter we use "deploy" to cover the process of making software available and "adopt" to refer to decisions by individual employees to begin using the software.

However it is labelled, the distinction is important. We argue that management support is often critical in acquiring and deploying group-support applications, but adoption by users is often bottom-up and unaffected by management attitude. The "management mandate" to use collaboration technologies is often not forthcoming and not a major factor in adoption. In the organisations we observed in the 1980s and 1990s, deployment of electronic calendars was widespread or universal, but adoption did not occur in the 1980s and did occur in the 1990s. This chapter focuses on adoption, the factors leading to initial use after deployment has occurred.

Because the technologies we are describing are primarily commercial software products, "success" and "failure" can refer either to commercial success or failure of a product (and thus its continued market existence or disappearance) or it can refer to the success or failure of a group's or organisation's particular effort to use it. The latter is often more difficult to determine, since a particular effort may be regarded by some as a success, by some as a failure, and others may perceive a mixed outcome, given that an application is rarely used precisely as envisioned. And these assessments can change over time.

Most group support applications and features have been commercial failures from the vendor perspective, but most are adopted and used satisfactorily by someone, somewhere. We use "success" and "failure" to refer to the usually less ambiguous commercial outcome.

By "collaboration technology" we focus on features and applications that support groups working together, such as chat systems, instant messaging, e-mail, electronic whiteboards, meeting support systems, shared calendars, document highlighting and annotation, application sharing, desktop conferencing, newsgroups, workflow systems and virtual worlds. The software may be widely available in an organisation, but it is adopted by groups to support their work together. The term "groupware" was used for the first stand-alone applications; today most software incorporates features to support communication, information sharing, and coordination, so the term "groupware" is losing currency even as the technology category has become more important.

8.1.3 Our Focus: Shared Calendar Use

Collaborative electronic calendar applications have had recent successes that stand in contrast to early, documented failures.

Not everyone keeps a calendar online and not everyone with an online calendar allows others to access it, but paper calendars are familiar to everyone. Familiarity with paper calendars can create special problems when people turn to electronic versions. People know what to do with paper calendars, which are generally private places to record professional and personal information. Electronic calendars move all that information into a potentially public realm. Although access to information can be controlled with privacy settings, doing so can require an effort to change or make exceptions to default settings, and changing privacy settings can conflict with the benefits of making calendars into collaborative objects. Facilitating the scheduling of appointments and meetings through shared access to information is a major appeal of calendaring applications.

Public use of a once private object makes the issue of discretionary or mandated use even more of a concern. How does an organisation encourage adoption of electronic calendars? Scheduling features are only useful with a critical mass of users, which argues for a mandate. However, calendar-keeping is an idiosyncratic practice and has always been discretionary. Can an inherently discretionary activity be mandated? Are there other trajectories for adoption?

About 20 years ago, commercial office systems began to provide meeting scheduling features with electronic calendar applications. Computer-assisted meeting scheduling represented an appealing technical solution to a clearly identified problem: people spend a lot of time arranging and rearranging meetings. Nevertheless, it was difficult to find successful use of the software, even in the development organisations and research laboratories that produced the software and provided widespread access to it. In fact, Grudin (1988) used meeting scheduling to illustrate factors that contribute to the lack of widespread adoption of some groupware applications.

Today, use is more widespread. Electronic meeting scheduling and calendaring occurs in environments similar to those where it did not a decade ago. In this chapter we review the findings of the mid-1980s, then describe two detailed interview studies conducted in environments much like those examined a decade earlier, in which online meeting scheduling had become routine. What changed, in the software or in the environments, that contributes to current use? What was the adoption process? The actual or potential presence in other settings of such factors, once identified, might signal the usefulness of this or other collaboration technology and suggest adoption processes to expect or encourage.

8.2 Early Computer-Supported Calendaring

By the early 1980s, many organisations had installed office systems that provided wordprocessing, spreadsheets, e-mail, and online calendars. In settings in which online calendars are widely available, the concept of computer-assisted meeting scheduling is simple: A person scheduling a meeting identifies the participants and the application checks each person's calendar, finds a time when everyone is free, and schedules a meeting or issues invitations. The potential utility was clear: the time spent scheduling and rescheduling meetings is substantial. Opportunities lost through inefficiency in meeting scheduling had been identified (Ehrlich, 1987a).

However, a study conducted by a company that had developed an early electronic calendar identified factors that contributed to a lack of use of the meeting scheduling feature (Ehrlich, 1987a, b; Grudin, 1988). Electronic calendars were used as communication devices by executives, managers and their secretaries, but only by about one in four individual contributors. The latter, if they kept any calendar at all, found portable paper calendars more congenial – available in meetings, for example. To maintain an online calendar would require more work of individual contributors, but the direct beneficiaries would be the managers and secretaries who called most meetings. In addition, although most employees had computer access, not all were networked tightly enough for the software to connect them. As a result, meetings were scheduled by traditional methods, despite the presence of the software on everyone's desks.

Research papers described new approaches to meeting scheduling that did not fare much better. Woitass (1990) designed a system to address the access problem. Based on a distributed agent architecture, it consulted the electronic calendars of those maintaining them and engaged other users directly. Although a technical success, Woitass reported that the system was not used for much the same reason: it required work from those who did not see the benefit of using it. Beard et al. (1990) developed a priority-based visual calendaring system that performed well in controlled experiments, but in field tests encountered a range of problems. Users were reluctant to use the publicly visible priority mechanism because it was easy to offend colleagues whose meetings were ranked less than high. Lack of integration with users' desktop applications also inhibited use. A 1991 survey of groupware use was sent to "empirical researchers, system developers, and end-users of CSCW (computer supported cooperative work) products and systems" (Butterfield et al., 1993). Respondents indicated that group calendaring systems were the most widely available of ten groupware technologies by a substantial margin, but were judged to be the least likely to have a significant impact.

Collaborative calendaring applications were not unique in requiring additional work (in this case calendar maintenance) from people who saw no benefit in their use. Most large systems and major applications face this challenge when introduced. To overcome this, management often hired support staff such as system administrators and data entry personnel, changed staff job descriptions, and encouraged or even mandated use. But collaboration technology is often not large or expensive – as with calendars, it can be a small application or just a feature in an individual productivity tool. In such cases the collective benefit is not as evident, it does not have the visibility that an expensive system does, so it is unlikely to get the same degree of management support. In sum, managers in the environments examined in the 1980s did not mandate that individual contributors maintain online calendars solely to facilitate meeting scheduling, nor were secretaries asked to maintain the online calendars of individual contributors.

Upper management advocacy is a key element in large system deployment and adoption. In contrast, the use of individual productivity tools or single-user applications has more often been discretionary; an application must offer each user a tangible reward. Collaboration technology is caught in the middle: it often must appeal to all group members, yet can expect little in the way of long-term, top-down advocacy. Collaboration technology such as calendar applications can benefit from managerially mandated *deployment*. This ensures that all members of the organisation know that their colleagues at least have access to the technology, which lowers the barrier to begin using its collaborative features. An organised roll-out can also eliminate incompatible competing applications, reducing the support burden for IS groups. However, although widespread deployment is an important step toward widespread adoption, access does not ensure use.

But the situation around calendar adoption was never deemed hopeless: "The conclusion is not entirely negative. Automatic meeting scheduling could be targeted to environments or groups making the most uniform use of electronic calendars. Their value can be enhanced by adding conference room and equipment scheduling..." (Grudin, 1988). "Focus on improving the interface for those who receive less direct benefit ... this may seem obvious but is very difficult" (Grudin and Poltrock, 1990).

It is difficult to focus interface design on "incidental" users because an application *must* appeal to its principal beneficiaries. A meeting scheduler that does not appeal to the manager or administrator who calls meetings has no chance. Interface development is usually directed toward those who will benefit most from using an application, not those who benefit least (Grudin, 1994b).

Ehrlich's (1987a) data suggested one way to target existing calendar users. She reported a higher incidence of secretary-supported calendar use among managers and executives in the development organisation she studied. Electronic support for scheduling meetings of managers was therefore feasible and has been reported. For example:

> PROFS is the single most used office system in the company. Most people with PROFS accounts do not use its calendar. Almost all managers/supervisors use the calendar, and secretaries maintain the calendar for 2nd level managers and higher ... They collected data in (a group of managers and secretaries) and found very large time savings and cost savings due to the online calendar." (Steven Poltrock, personal communication, 10 March 1994)

Obstacles hindered the wider use of this system in the company:

> In my [group] most people use an online calendar, but our calendars are not shared – we are not using PROFS. (ibid.)

Several years after the early meeting scheduling features were marketed with little success, a new generation of calendar applications began to be widely adopted in engineering environments very similar to those which had rejected them previously.

8.3 Computer-Supported Calendaring in the Mid 1990s

Electronic calendar applications that appeared in the early and mid 1990s were much more sophisticated than those from a decade earlier. They began to include more features to support the private, idiosyncratic information management aspects of calendars (albeit less than perfectly). These electronic calendars predate the advent of mobile computing, so adoption still faced the obstacle of being mostly anchored to desktops, precluding electronic access outside the office and work hours.

In some workplaces, notably engineering environments, use of shared electronic calendars was gaining ground. Use elsewhere was far from commonplace, but it made sense to study the early adopters, especially because it was precisely in such engineering environments that calendars had not made headway ten years earlier. Our research questions included:

- How widespread is calendaring and meeting scheduling at these sites?
- Who uses them?
- Are new application functionality or interfaces factors in adoption?
- Has the collective benefit of use been made more salient?
- Is adoption the result of top-down encouragement or mandate?
- How does adoption proceed over time?

8.3.1 Method

During a three month internship in a Microsoft development division, the first author observed and used scheduling activities in her day-to-day work. During this time, she also conducted five formal interviews on the topic of meeting scheduling. This experiential and interview data informed a set of 40 questions used in twelve interviews conducted over a three-day period at Sun Microsystems. We had learned of widespread use of electronic meeting scheduling at these sites and sought active users, as well as some non-users. Our interest was in identifying possible factors that led to such high adoption rates; the specific adoption contrast that frames this paper emerged during the analysis.

Later, based on the findings from these two sets of interviews, we designed site-appropriate surveys comprised of 20 questions that were administered to about 3000 people by e-mail at each of the sites. These surveys were intended to validate our interview findings and extend the inquiry to employees across different business divisions of the corporations.

Our interviews and surveys queried subjects about their use of paper and electronic calendars. Questions covered procedures and protocols for arranging or responding to meeting invitations, privacy, granting calendar viewing access, peer pressure, managing one's calendar when away from the computer, using calendars for structuring work, and using calendars for handling resources such as conference rooms and equipment.

8.3.2 Site, Participant and System Descriptions

Microsoft Corporation

At the time of this investigation in 1994–95, most of Microsoft's then 15,000 employees were located in the Seattle area, where the interviews were conducted. The average employee age was 34, low for a large software company. For many, Microsoft was their sole or principal work experience. Most employees had their own office or shared with one other person, and every employee had ready access to a common platform computer that was linked to a company-wide network.

The five interviews were conducted with people in different job positions and in different workgroups across the development divisions. All had been employed at Microsoft before the electronic calendar application was introduced, with an average of about 5 years' service. Listed in increasing order of time spent in meetings, they were a developer, visual designer, program manager (project leader), development instructor and first-level manager.

The follow-up online survey was administered to approximately 3000 employees over distribution lists, with a response rate of approximately 30 per cent: 47 per cent of respondents were software developers and testers; 20 per cent were project leaders or first-level managers; 8 per cent were technical writers and visual designers; 7 per cent were marketers; 4 per cent were mid- or executive-level management; 4 per cent were interns; 3 per cent were administrative assistants; with the remaining 7 per cent in other positions.

Their calendar application used at the time of this investigation was Microsoft's Schedule+ running on Windows PCs. With Schedule+, calendars resided on the client machine with frequent updates to a central server. Users viewed a calendar by specifying its owner's e-mail alias or name. Schedule+ was developed in-house but was a commercial product, intended to support meeting scheduling in a range of organisational cultures. All employees could download the application from a server.

Sun Microsystems

At the time of our investigation, about half of Sun Microsystems' then 13,000 employees worked in the San Francisco Bay Area, with about 3000 in the engineering departments where our interviews took place. The average age of Sun employees was 38. Most employees had private offices or shared with one person, except for administrative assistants ("admins") who were typically in centrally located cubicles. Employees had workstations connected to a company-wide network.

Eleven electronic calendar users from ten engineering groups and one non-user in corporate marketing were interviewed. They had been at Sun over five years on average; two arrived after the electronic calendar was introduced. Three participants were or had been on the Calendar Manager product team and provided technical implementation information, an historical account of the product's introduction, and data about their personal use. Listed in roughly increasing order of time spent in meetings: two administrative personnel, two technical writers, three developers, two human interface designers, one product marketer, one manager, and one corporate marketer.

The follow-up online survey was administered to approximately 3000 employees over distribution lists, with a response rate of approximately 50 per cent: 22 per cent were developers or testers; approximately 20 per cent were engineers; 13 per cent were first-level managers; 9 per cent were mid- or executive-level managers, 9 per cent were administrative assistants; 7 per cent were marketers; 4 per cent were salespeople; 4 per cent were customer support representatives; 3 per cent were technical writers and user-interface designers; 3 per cent were system administrators; 3 per cent in business administration support; with the remaining 3 per cent in other positions. (The number of engineers is approximate because a survey omission required respondents to use the "Other" category and manually write in their job positions.)

At the time of our investigation, Calendar Manager ran under OpenWindows and CDE on SparcStations, and was available as a standard desktop tool. Calendars were maintained on their owner's host machines. Viewing other calendars required specifying user names *and* the host machines, a deliberate action that seemed to reduce casual "surfing" of calendars. Host names, however, were listed in an online corporate rolodex of employee information which made browsing across the corporation possible.

8.3.3 Description of Use

Adoption Levels

Adoption of the electronic calendar applications was widespread at both sites. Since we had sought informants who used online meeting scheduling for the interview portion of the data collection, it was not surprising that high usage was reported in 16 of the 17 interviews (i.e., 15 of the 16 groups). The exception was the Sun corporate marketing representative, a former user, who frequently met with external contacts, people not accessible through the system.

Survey data revealed more information about adoption levels. Although we encouraged users and non-users alike to complete the survey, we assume that non-users were less likely to respond. For that reason, we do not estimate adoption levels based on user/non-user responses. Instead, we asked respondents to estimate how many people with whom they worked used the electronic calendar. Based on these responses, we conservatively estimate adoption to have been at least 75 per cent at both sites at the time of investigation.

Furthermore, users tended to use the application regularly and keep their calendars highly up-to-date. Specifically, 81 per cent of Sun survey respondents said that they recorded *at least* three-quarters of their appointments in their calendars; 75 per cent of the Microsoft respondents said the same.

Despite the widespread adoption of the calendar applications across the two corporations, their extensive use was not perceived as remarkable by employees. The calendar applications existed below the horizon of notice, perceived simply as tools in the conduct of mundane activities of work. However, electronic calendars were clearly important to employees, and the integration of electronic calendars into everyday work was essentially complete: When queried, interview informants could not imagine life without them.

Privacy Setting Defaults

Electronic calendars are surprisingly diverse in the kinds of collaborative uses they afford. Privacy setting defaults seem to significantly dictate how an electronic calendar application can be used socially: users tend to maintain the defaults and a culture of use grows around that deployment decision, further reinforcing adoption of the default settings (Palen, 1998, 1999). In single-user applications, users tend not to change default settings (Mackay, 1991). As the cases of Microsoft and Sun illustrate, the same tenet applies to collaboration technology, even in high-tech environments with knowledgeable users, affecting use on a large scale.

Survey results indicate that 81 per cent of Microsoft's Schedule+ users and 82 per cent of Sun's Calendar Manager users maintained their privacy setting defaults. However, the default privacy settings were different between the two applications (and remain different over five years later). At Microsoft, the default settings restricted the amount of calendar information that could be shared with others: only free and busy times could be viewed, with no other appointment detail available. We refer to this as a Restricted Access model calendar.

In contrast, Sun's Calendar Manager is an Open Access model because its privacy setting defaults would allow everyone on the network to read the full contents of others' calendars, including appointment time, appointment description and even any additional notes entered by users. Descriptions of private appointments, such as job interviews and medical appointments, must therefore be individually locked.

Determining Availability and Scheduling Meetings

The differences between the default settings had remarkable consequences on the applications' uses in practice.

With so much information available about their colleagues' activities, Sun employees used Calendar Manager not only to schedule meetings, but also for general coordination activities. Participants explained that being able to see the content of calendars was very useful: they could infer whether colleagues are in their offices, or learn where they are and when they might be back, if they could be interrupted, or which planned meeting might be more easily rescheduled in case of a conflict, and so forth. For example, colleagues could infer that a meeting that occurs across town requires a half-hour commute, even if that time is not allocated in the calendar. The ability to make these inferences can lead to more efficient scheduling than a built-in "auto-pick" feature. We found that the auto-pick feature has a place in scheduling very large meetings, but automatic meeting scheduling was otherwise not a part of normal scheduling practices.

People unfamiliar with Open Access calendar environments often wonder whether it might lead to abusive surveillance or "micro-management". In practice, the coordination benefits just described are so great that abuses rarely or never occur. The value would disappear if harsh practices made people reluctant to maintain their calendars. In fact, subsequent research on Sun's use of their Open Access model calendar system revealed that Calendar Manager was (and still is) used as a kind of distributed information system that significantly supported coordination beyond meeting scheduling. These findings are presented in Palen (1998, 1999).

To initiate a group meeting, Sun employees would typically send regular e-mails listing a possible time after viewing colleagues' calendars. Many were either unaware of the various ways the calendar system can generate a "template" that recipients can drag-and-drop into their own calendars, or they found them hard to generate. The ease with which people could infer colleagues' availability with the open calendars, combined with the work required to use these meeting templates, meant that Sun employees were more likely than Microsoft employees to bypass the system to set up a small meeting. However, Sun employees generally entered such appointments in their calendars after scheduling them.

In contrast, at Microsoft, employees would typically schedule a meeting by checking the availability of others and sending Schedule+ mail with a proposed time, allowing the recipients to accept or decline with a keystroke. This accepted approach, with no informal prior negotiation of time, can strike outsiders as very blunt when it involves, for example, initiating a one-on-one meeting with someone one does not know and who is not expecting the meeting request. Because no additional information other than when someone is free or busy is available, employees would take more of a hit-or-miss approach to finding an agreeable time.

Planning a meeting when away from the office required the evolution of special scheduling practices. Since this investigation predated mobile technology, calendars could only be viewed on desktop computers. Given this context, surprisingly few people reported making hardcopies of their calendars. In fact, only 32 per cent of Sun respondents and 13 per cent of Microsoft respondents reported that the printing feature was of value to them.

Instead of relying on hardcopies to schedule meetings when away from their desktop machines, users would tell colleagues to wait until returning to their offices. A short-hand language emerged in support of this practice: At Microsoft, phrases like "Sched plus me" or "Plus me" were commonly used. At Sun, the feature name for viewing calendars was adopted, so phrases like "Browse my calendar" or "Browse me" were often heard in response to a meeting request.

Resource Scheduling

Both sites used the calendar applications to schedule resources, although Schedule+ supported conference room scheduling more systematically than Calendar Manager. At Microsoft, all conference rooms were allocated their own calendars, allowing the conference rooms to be "invited" to meetings.

A competing calendar application at Sun, SchedRoom, kept conference room scheduling out of the domain of Calendar Manager, essentially requiring users to schedule a meeting using both applications (for more discussion on the co-evolution of these two applications, see Palen, 1998). However, Sun employees used Calendar Manager for scheduling other kinds of shared information, including groups' vacation schedules.

8.4 Factors Contributing to Successful Adoption

Interviews and surveys, and inferences drawn from them, provide only a partial view of the adoption process: they rely on retrospection and memory of practices. That said, a pattern of discretionary adoption and a set of factors contributing to it did emerge that differed from those prevailing ten years earlier in similar environments. In summary these include:

- changes in infrastructure, including networking, support, and behaviour;
- expanded application functionality;
- improved graphical interfaces that provide versatility and ease of access.

With respect to the central question of how groups reach a critical mass of use, we found that a high-level management mandate was not a factor. In a few cases a group leader strongly encouraged group members to adopt. We interviewed one self-described calendar evangelist. Most often users reported bottom-up adoption abetted by peer pressure. Peer pressure had technical and behavioural aspects:

- features and functions of calendar software, notably its integration with e-mail, that remind non-users of use by others and benefits they might be missing;
- a bottom-up pattern of adoption, with use spreading from individual contributors to managers and administrative assistants ("admins"). Pressure on holdouts, sometimes mild and sometimes less so, could eventually come from every direction – managers, admins and peers – and adoption became nearly universal.

In this section we detail these five observations.

8.4.1 Infrastructure

Our principal sites, Sun and Microsoft, had organisation-wide platforms supporting the same calendar and scheduling software to which all employees were networked. In other sites we examined, users could exchange e-mail over a corporate network, but sometimes had incompatible calendars.

Ten years earlier, this was not true in large high-tech companies. Employees were variously supported by corporate mainframes, divisional minicomputers, and small clusters of PCs that were difficult to network. E-mail did not span these environments, much less meeting scheduling software. Members of a team often shared a platform, but many meetings included people from other groups; therefore the software could not routinely replace traditional scheduling methods. Francik et al. (1991) identified the fact that working group boundaries often do not match organisational chart units as a key obstacle to adoption.

In our study, our one interview subject who did not use an online calendar in corporate marketing had many of her meetings with people external to the company whom the software could not reach. Ten years earlier, most developers were in comparable positions with respect to other groups *within* the company.

Both sites also provided strong technical support in installing and maintaining the software. This arose in several interviews. Group calendaring over a large client–server network is technically challenging. The failure of two releases of Calendar Manager was attributed by one informant to reliability problems, and occasional losses persisted:

> About 2 years ago [sigh/laugh], I lost all my data. And they had no idea what happened, it just was gone. So I mean all my personal data was gone too ... I felt so stupid that I had to go back and call all these people. I lost all my birthday data ... my work data ... I was lost, because I rely on my calendar to tell me where I am supposed to be and what I am supposed to be doing. I really was relying on it. When I lost the data, I had to go around and try to find it, or I'd get calls "where are you?" "well I am right here, where am I supposed to be?" So, you know it was an embarrassing time.

Clearly not much of this will be tolerated in a product.

Finally, although difficult to measure, behaviour has changed. People are much more comfortable with technology and are heavier users of it now than was true ten years ago, even in engineering environments. Almost everyone in these settings could be counted on to read e-mail regularly, one element of what Markus (1987) calls "message discipline", an important factor in promoting use. The casual use of phrases such as "Sched plus me" and "Browse me" is further evidence of structuration or an evolution of the culture around the technology. A "behavioural infrastructure" was also in place.

8.4.2 Functionality

Another change over the ten years is that the products matured, supporting many new functions. Unlike earlier products and research systems, these applications were strongly integrated with desktop environments, e-mail and corporate online rolodexes. The tight coupling with e-mail was singled out as a factor in early adoption. For example, 74 per cent of the Microsoft respondents said that receiving Schedule+ e-mail played a role in their decision to start using the system. Notably, 88 per cent of student interns cited Schedule+ mail as a factor in their adoption of the system. Interns have little time to socialise to the organisational culture, and their experiences indicate how influential the technologically abetted peer pressure could be.

Conference room availability was sometimes described as the most critical aspect of scheduling a meeting, so the ability to schedule rooms or equipment through the system is another example of a feature that promotes use. At Microsoft, where the only way to schedule a conference room was through Schedule+, 57 per cent of all respondents cited this functionality as a reason for adoption. This feature was even more important to populations that schedule meetings frequently, such as group and first-level managers, and the administrative assistants who schedule meetings on behalf of higher-level management: 80 per cent of admins and 77 per cent of group and first-level managers reported that the conference room scheduling feature was a very important adoption factor. (At Sun, a different system was used for most conference room scheduling, although when administrative assistants wanted control over a conference room, they "moved" the conference room to a Calendar Manager calendar, where only those who knew about its existence could schedule the room. For them, Calendar Manager was a useful tool for wielding control.)

Two seemingly minor features cited as most important by survey respondents were the diverse mechanisms for reminding oneself of an upcoming event and more flexible ways to define recurring meetings, such as "Mondays, Wednesdays and Fridays at 9". These features support individual work practice but proved important in achieving a level of adoption that allowed the application to support collaboration. In fact, at Microsoft, automatic meeting reminders appealed more to individual contributors than to other employee categories, considered important by 93 per cent of developers but only 60 per cent of admins and 70–75 per cent of managers of different levels. (The data by job position is not available with the Sun survey data, although interview data suggests similar trends.)

At Sun the ability to view other people's calendars was roughly equal in importance to these two individual productivity features. Although Microsoft employees felt that viewing other calendars was less important, they reported that value of using the system came when other people also used it. This difference may be accounted for by the way calendars were used at each site: at Microsoft, viewing others' calendars was important for scheduling meetings, but little other information could be viewed to apply to other purposes. Sun employees' use of others' calendars, on the other hand, extended beyond meeting scheduling to support other kinds of coordination.

Both systems provided a broad range of privacy options. Informant attitudes toward privacy differed, so this flexibility is important, especially in environments that use technology that supports a high degree of information sharing. In the Sun case in particular, where the calendar is deployed with open default privacy settings, we found that the option to change settings was important, even if people did not act on the option. Giving total privacy control to users further encourages use of the collaborative technology.

One admin who had used Calendar Manager for six years to keep calendars for managers, but only started keeping her own for the last three, said:

[I started using it for myself] when I realised all the things it could do ... I don't remember it being that sophisticated before ... I mean it is an incredible thing, I mean now you can show multiple calendars on top of each other. So it's really just gotten so good, it's become a valuable scheduling tool ... I would say that with all of the increased capabilities of Calendar Manager that a lot more people have started to use it and realised that it's a very helpful thing.

8.4.3 Versatility and Ease of Access

In discretionary use situations, ease and enjoyment are powerful motivators. With the applications studied, many calendar and scheduling features were easily accessed and many tasks could be carried out multiple ways, with one or another way being more natural depending on the context. With the earlier generation of schedulers, obtaining information was too time-consuming to promote some of the casual uses of calendar information that we observed. Even when unrelated to meeting scheduling, versatile access and management of calendar information promoted the prerequisite online calendar use; they contributed to "calendar discipline", the behavioural infrastructure.

We found evidence that for applications that are only indirectly tied to people's principal work missions, interface transparency and efficiency were particularly important. Where a feature is even slightly obstructed, it can go unused. For example, Calendar Manager users greatly appreciated receiving invitations in the form of a "meeting template" which could then be dropped into their calendars. Yet, only 26 per cent of the Sun survey respondents reported sending meeting templates regularly. One survey respondent reported: "I love it when people send the attachment that I can drag and drop, but I don't do it (I haven't figured out how to do it)." Furthermore, many message senders who did send templates did not use either of the ways to drop a meeting template into their invitation, each of which involves an extra step or two. Instead they took even more time to type an invitation character by character to conform to the message template. Another survey respondent explained: "I would do it more if the template was automatically generated for me."

One developer pithily summarised his attitude: "walk arounds are work arounds". If he felt blocked by the application, rather than looking for a way to work around the problem, he would arrange meetings "off-line", on foot.

An interface change that was widely praised at Microsoft was allowing users to enter employee's actual names, in contrast to their e-mail aliases. At Sun, employees' calendars had to be identified by name and host machine. The integrated online rolodex rendered this only a minor inconvenience; nevertheless, this was a common source of complaints, suggesting that small interface details are salient to users who may not be strongly committed to an application.

8.4.4 Universal Adoption and Peer Pressure

> The only way that makes it useful is that everyone is using it.
> (Product marketing representative)

Use was almost universal in the groups sampled, with some people unaware of anyone who did not use it. Across the corporations, adoption was conservatively estimated at 75 per cent, as indicated by survey results. A Sun developer remarked:

> The numbers have gone up dramatically. I think now ... everybody uses it, anybody who has a workstation uses it. It's hard to find someone who doesn't use it ... It'd probably be the janitor or someone like that because they just don't have access to the equipment. Everybody ... even non-technical people use it.

How did use become so widespread? When asked directly, some reported that they felt peer pressure to keep calendars online – 50 per cent of the Sun respondents and 59 per cent of the Microsoft respondents said they felt at least some peer pressure to use the system. However, many who reported feeling no pressure said that was because *they* were the ones in their workgroups to deliver pressure:

> I pressure managers "considerably" to use their calendar manager! Every so often I end up with one who just WILL NOT use the tool, and it drives me nuts!

Still others felt that "pressure" was too strong a word to describe the expectation they felt from others to use the calendar system. A Sun survey respondent wrote:

> I don't feel that there is explicit pressure, just a widespread expectation that everyone is using it. I am one of the people who raise eyebrows when I find out someone is not using it and it affects my work. I've stopped short of requiring each of my staff to use it.

Similarly, interview informants who said they did not notice pressure often contradicted themselves, in subtle or not so subtle ways, by also reporting clear frustration with non-users. For example, a Sun admin reported telling the only non-user he knows,

> Everyone on the team is available at this date and time, except I don't know whether you are even in your office.

Similarly, a Sun human factors engineer who said that he didn't believe there was any pressure to use the calendar nevertheless expressed annoyance with non-users, a sentiment that is probably hard to disguise:

> Where I find that things become annoying is that a lot of people will call for meetings without using the calendar appointment embedded in their message ... then I've got to go into the Calendar Manager and schedule an appointment and it seems like a waste of time. And that happens all the time.

In environments where "sched plus me" or "browse me" were common phrases to delay the scheduling of future meetings until users were back at their computers, non-users would quickly feel left out. In fact, non-users often missed meetings entirely that were scheduled without their notice. Furthermore, lists of "appointments" that included company holiday dates and colleagues' birthdays that could be automatically inserted into electronic calendars further encouraged participation.

Finally, the application design can itself contribute to social pressure. Recognisable appointment attachments reminded non-users of a benefit they may be missing. Calendar integration with e-mail enabled users to deliver peer pressure to use the calendar easily via the technology. As a manager at Microsoft reported:

> Some people don't look at my Schedule+ [calendar] to see if I'm available. I will press decline, and say, "Please see my Schedule+."

The recipient was left with little choice other than to use the software.

8.4.5 The Adoption Trajectory

No single adoption pattern fits every group. A few interview informants reported pressure from a manager or project leader. But more typical were those who reported that developers – individual contributors – were the first adopters, exerting pressure on peers and then on managers and admins, with pressure continuing up the managerial ladder. Once admins adopted meeting scheduling they could exert pressure in all directions. (It may be significant that admins now appear to wield more influence than did secretaries in the 1980s.)

People expressed a variety of perceptions about adoption. A Microsoft program manager said: "I resisted for six months. When I had people under me, I needed it. Now I can't live without it." A Sun corporate marketing representative, who initially used Calendar Manager but has since stopped, reported:

> [In] Sun culture, engineers [say] "You are not on Calendar Manager?" They're just shocked that you don't use your own technology ... I don't think marketing management uses [Calendar Manager].

A Microsoft trainer reported that "the first groups to get up to speed were the admins". Similarly, a Sun product marketing representative said:

> I think that most people keep a Calendar Manager because of the browsing capabilities ... When we are trying to set up appointments, admins ... rely on being able to look at other people's calendars to see when they are available.

A Sun developer described a bottom-up adoption pattern:

> Over time, they (managers) started to use it, they started seeing the benefit of using it ... but initially I can't say there were a lot of people using it. Initially ... it was mostly developers, then you start getting a scattering of managers, I bet you now we could go and browse Scott McNeely's (Sun CEO) calendar if we wanted to, so now all the way to the top ... now everybody.

However, a Sun technical writer reported a different experience: "pressure was from management, not coworkers. The attitude is 'Get into the 90s.'"

Survey results support a largely bottom-up adoption trajectory: At Sun, 51 per cent of the respondents felt that their immediate colleagues were a source of peer pressure, whereas only 29 per cent reported feeling pressure from their management and 25 per cent from admins. At Microsoft, 70 per cent of the respondents said that pressure came from their colleagues, with 25 per cent reporting pressure from management, and 7 per cent from admins. Focusing on high-level management, 80 per cent of Microsoft's managerial respondents reported feeling pressure from their colleagues, and 40 per cent reported feeling pressure from their staff members. Only 23 per cent of managers reported feeling pressure from *their* management.

8.5 Conclusion

We examined organisations in which calendaring technology was acquired by developing it and was very widely deployed as a natural consequence, although these decisions required high-level management involvement (see Palen, 1998, 1999, for a discussion of Sun's development and deployment over a decade). However, deployment does not necessarily lead to adoption by employees, as we have noted.

Despite arguments that only mandated use of online calendars will ensure the critical mass of use needed to facilitate meeting scheduling, this did not occur. Instead, a new generation of calendars was more useful and easier to use, supporting discretionary adoption. The calendars could be shared with virtually anyone in these organisations, and by the 1990s almost all employees used desktop systems.

In the 1980s, individual contributors did not keep calendars online. The applications of the 1990s added features that appealed to individual contributors, such as meeting reminders and invitations or appointment

attachments that are easily dropped into a calendar. Integration with e-mail meant that non-users received frequent reminders that the software was being used by others.

Instead of top-down, mandated decrees to adopt, we saw a common adoption trajectory: some individual contributors chose to use the software to increase their individual productivity. Once enough people did so, the potential benefit of shared calendars became more evident. As people grew accustomed to using calendar applications to coordinate, the relative difficulty of scheduling non-users stood out, leading to peer pressure to adopt.

Initially, the collective benefit of easier scheduling translated into more work for individual contributors and more benefit for the managers and secretaries or admins who organise most meetings. The solution was not to mandate that the individual contributors do more work; rather, it was to reduce the work required of them by building better interfaces and features that they did value. This clearly seems desirable in principle, and in practice it was necessary because few organisations were ready to mandate the use of such a small, personal application as a calendar.

The use of collaboration technology may not always follow this pattern. For one thing, system-wide deployment may be a necessary prerequisite – a high-level management role can be important here. Vendors once hoped that groupware would be deployed one group at a time, but this has proven problematic due to the dynamic nature of workgroups, with membership distributed organisationally and changing over time.

However, factors that plausibly affect adoption of collaboration tools, such as their decreasing cost and visibility, have grown stronger. The core observations are likely to be general. For example, a recent study of the widespread use of NetMeeting within Boeing concluded "it is generally the users, and not management, who are the driving force in diffusing the technology across distance" (Mark and Poltrock, 2001).

Another path to adoption is seen in new enterprises that organise around collaboration technologies. Their employees form interaction conventions based on the software, rather than having to give up or change conventions established previously.

How will the observations of calendar use generalise to other collaboration technologies? With collaboration features proliferating and the cost of applications decreasing, most will have the low organisational visibility that suggests that widespread deployment followed by bottom-up adoption is likely to be the norm. This in turn will require features or processes of use that deliver benefits to most or all group members.

Acknowledgements

This chapter extends work published in Grudin and Palen (1995). We thank interview participants at both sites; Ellen Isaacs for her enthusiasm and intellectual and operational support; Rick Levenson, Mark Simpson, Susan Denning, Kent Sullivan and the Microsoft Usability Group for their support; and Susan Ehrlich Rudman, Steven E. Poltrock, and Lynne Markus for comments and help identifying details of calendar use. This work was supported in part by a National Science Graduate Fellowship and National Science Foundation Grants #IIS-9612355 and #IIS-9977952.

9

Supporting the Assimilation of Sector-Specific Collaborative Solutions: Ten Commandments

Bente Evjemo, Sigmund Akselsen and Jan Grav

9.1 Introduction

Different groups and organisations have gradually taken up collaborative solutions. The general impression is, however, that the great success stories are few and that companies and organisations find it hard to gain – or the problem might as well be a matter of identifying – significant positive effects. So, what factors are decisive for successful adaptation and transfer of collaborative solutions?

The answer may implicitly lie in the following three assumptions on how technology might increase a group's efficiency (McGrath and Hollingshead, 1994):

- *Increase a group's efficiency with regard to problem solving* by streamlining and structuring internal group communications and supporting/making efficient the method used to complete a task.
- *Remove the traditional limitations of time and place created around group activities.* Offering full, fast and interactive communication between participants divided in time and space, and thereby making it possible to establish work groups in previously impossible ways.
- *Improve a group's access to information.* By using technology, individuals and groups are given faster access to greater quantities of information. The technology can also facilitate better quality of information and the ability to process more.

These bullets might very well describe the overall vision of Telenor R&D, from where the following examples are collected. Telenor is a leading ICT company in Norway (see www.telenor.com), with several R&D units addressing various aspects of telecommunications. The unit located in Tromsø, a small city in northern Norway, counts about 25 persons. They represent all together a wide range of disciplines: telecom engineering, computer and information science, sociology, political science, economics and psychology. Through a multidisciplinary approach we have developed and introduced ICT solutions for selected sectors and, in addition, established a sociological and economical understanding of the demand and opportunities for ICT solutions in various professional settings. Our projects have to some extent addressed challenges typical for rural and scarcely populated regions, and have somehow included technological support to reduce negative effects of long distances between distributed co-workers, teams or companies.

We have been practitioners within this field for more than ten years, and have on the basis of our experience outlined some practically oriented guidelines on how to introduce collaborative solutions to various teams within different sectors. Working in Telenor's research department we are expected to produce knowledge, but being part of a commercial telecom company we are directed by consultancy and product generating expectations as well. This dualism influences on the research approach and consequently on the results. In most projects we try to merge academic and practically oriented interests.

Collaborative solutions are typically designed with one or more of the above assumptions in mind and diverse from the wide range of ICT solutions in several ways. The following two statements are seen as most important:

1. collaborative solutions are complex and closely bound up with a group's working methods, *and*
2. collaborative solutions must reach a critical mass before they become useful.

These characteristics have consequences for the adaptation process as well as for the transfer process. Grudin (1989, 1994b) discusses how these aspects have an influence on the systems developer's work, and presents results which we find familiar as well as inspiring.

Through close collaboration with workers in different sectors we have found that collaborative routines are not easily identified. This implies that the optimal features of a collaboration solution or the adaptation strategies are far from obvious. Our initial starting point might be one of these:

- We focus on some general challenges related to management or working routines in professional organisations.

- We find ICT products or concepts of special interest and look for areas for applications.
- We investigate particular sectors or companies for the sake of identifying their existing or coming need for supportive ICT.

Most often these initiatives are done in parallel. Nevertheless, to achieve our main goals we need discussion partners and test beds. Thus, our trademark has been case studies and close user involvement. According to Yin (1994, p. 13), case studies can be described as "an empirical inquiry that: investigates a contemporary phenomenon within its real-life context, especially when the boundaries between phenomenon and context are not evident". This is an all-embracing work style, involving both quantitative and qualitative approaches.

Our methods include some aspects of case studies, but are better characterised as action research where the unit for study (user group) also actively participates in the research process (Whyte, 1990). This type of action research – Participatory Action Research (PAR) – has its origin in Norwegian work union research (Thorsrud, 1977; Elden 1979) and reflects a Scandinavian tradition that the importance of a common learning process between researchers and organisation representatives should be emphasised (Elden and Levin, 1990). PAR provides an arena where participants, users and researchers exchange information and experiences, and make well-informed choices, including the decision to participate (Argyris and Schön, 1990). This is a cyclical investigative process, which includes identification of problems, action planning, implementation and evaluation of results. The goal is to produce solutions for the existing problem, and also to enhance existing knowledge of the special type of problem which is being investigated (Elden and Chisholm, 1993).

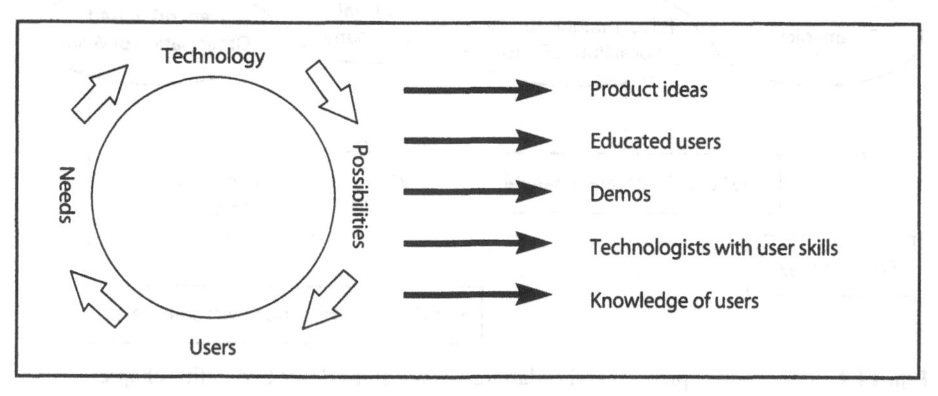

Figure 9.1 Results from mutual influences between technology and users – possibilities and needs (Akselsen et al., 1996).

This approach leads you into a cycle where the apposition of technological possibilities and users' needs fires a development – a process that might be fruitful and give different kinds of output: product ideas, skilled users, technologists with detailed sector relevant knowledge, and knowledge about users (see Figure 9.1). These results are valuable input to telecom businesses (Stenvold et al., 1997), and from a researcher's perspective they are candidates for further academic investigations as well.

In the following we will summarise what we have actually learnt through the last ten years of adaptation and transfer of collaborative solutions. What kind of organisational structures have we visited and what are the main lessons?

9.2 Overview of Projects

Through the 1990s we ran four main projects (Figure 9.2): *Telemedicine* (1988–94), *Telecommunication within the Local Public Sector* (1993–95), *Virtual Company* (1993–95) and *Network-based Organisation of Work* (1995–98).

These projects all focused on how to exploit ICT at its best, but encountered particular challenges within different sectors or professional organisations, where users had various technology experience and the teams were of different sizes. The technological solutions involved were different as well. The units of the companies involved were typically small or medium sized (from around 5 to 50 people) and in the position of developing their own strategies.

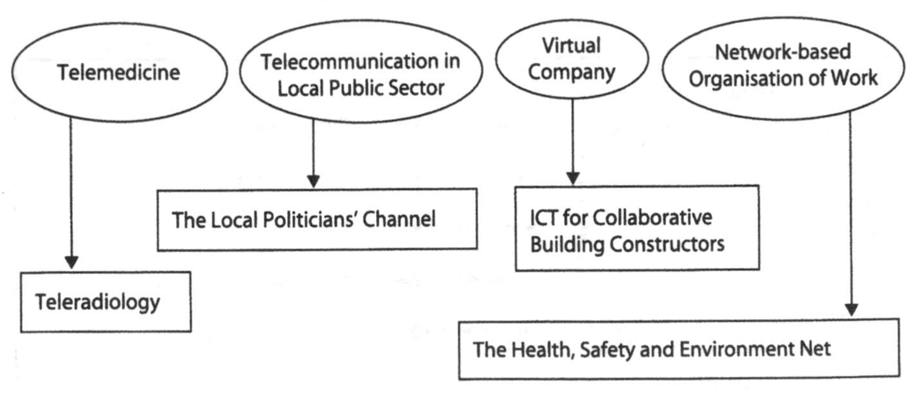

Figure 9.2 Overview of projects and collaborative solutions described in this chapter.

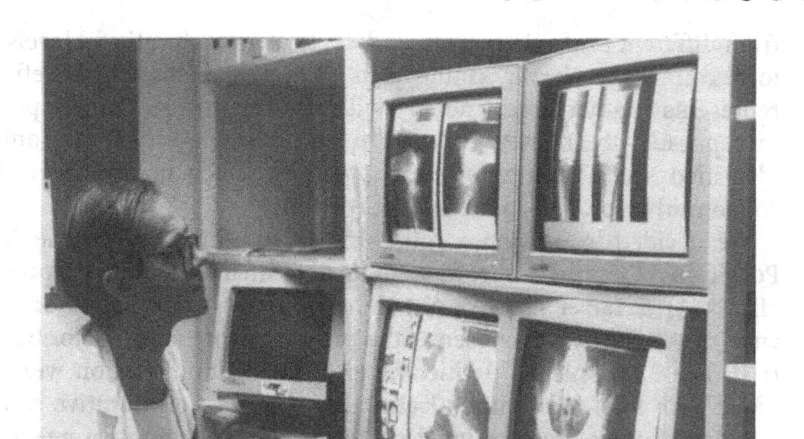

Figure 9.3 Radiologist diagnosing images sent from small hospital.

9.2.1 Teleradiology

The development of a Teleradiology solution (Sund et al., 1991) was one of several initiatives done within the Health Care Sector (Nymo, 1993). This trial had, as most of our telemedicine activities, the purpose of making health information available wherever needed and to avoid expensive, cumbersome and time-consuming transfer of patients and health personnel. In this particular case the origin was a small regional hospital in a rural area of northern Norway. It had X-ray equipment and specially trained nurses, but no radiologist at the site. The radiology service was provided by specialists from the University Hospital in Tromsø driving back and forth once a week. This was costly in both time and money. We developed a solution to digitalise and transfer X-ray images to the radiologist's workstation at the hospital, whereby the images could be diagnosed immediately, without the time consuming travelling and without any delay for the patients (Figure 9.3). Technically this became a success, and the small hospital had its X-rays analysed on a regular basis every day.

9.2.2 Local Politicians' Channel

The attention paid to the local politicians was part of our efforts towards the Local Public Sector in general where we ran several field studies within education, communications, utility service provision, etc. The local politicians' situation was special in the sense of gathering people from different professions, ages, and societal and educational levels. The local councils in Norway count from 15 to 85 persons, which definitely represents a rather large group of collaborating people. As these persons met sporadically and were engaged in politics in their spare time only, we identified special communication patterns, which no commercialised collaborative solution could meet at the time.

We decided to develop a collaborative solution denoted the "Local Politicians' Channel", introduce it to a council and evaluate potential effects (Ytterstad et al., 1996; Watson et al., 1999). A northern Norwegian rural municipality was chosen as the project partner and site for the field trial. The 33 municipal politicians and their administration were provided with PCs and communication lines, and a collaborative solution adapted to their organisational structure and collaborative needs. The system's highly graphical interface made it easy for politicians to make phone calls, set up telephone conferences, send and receive e-mail and faxes. An integrated document handler supported the exchange of documents between politicians and local government officers. The municipal council was followed up and supported throughout the project period of two years. In the effort to develop this solution we met the challenge of making collaborative systems for "the average man"; there were taxi drivers, high school teachers, nurses and farmers within the same group. We also experienced the tension of political power and how this aspect influenced the adoption of the system. The council did not continue to use the system after the termination of the project. The official explanation was lack of economical benefit (Watson et al., 1999).

9.2.3 The HSE Net: A Collaborative Solution for a Distributed Health, Safety and Environment Team

In times where flexible and dynamic organisational structures emerged, we turned to the special challenges of management and collaboration in distributed teams and how communication technology might facilitate these working arrangements. The project "Network-based Organisation of Work" aimed at producing knowledge that could be useful to companies in their choice of ICT solutions for future organisation of work (Akselsen, 1999). Results from the project can be found at http://www.tft.tele.no/netto. One practical part of the project was concentrated within Telenor's Health, Safety and Environment department, which was about to reorganise their way of working. We developed and implemented a collaborative system for a distributed team (Evjemo et al., 1999), focusing on management, daily collaboration and team maintenance.

The overall purpose of the solution is most easily explained through a description of how collaborative routines and relationships develop

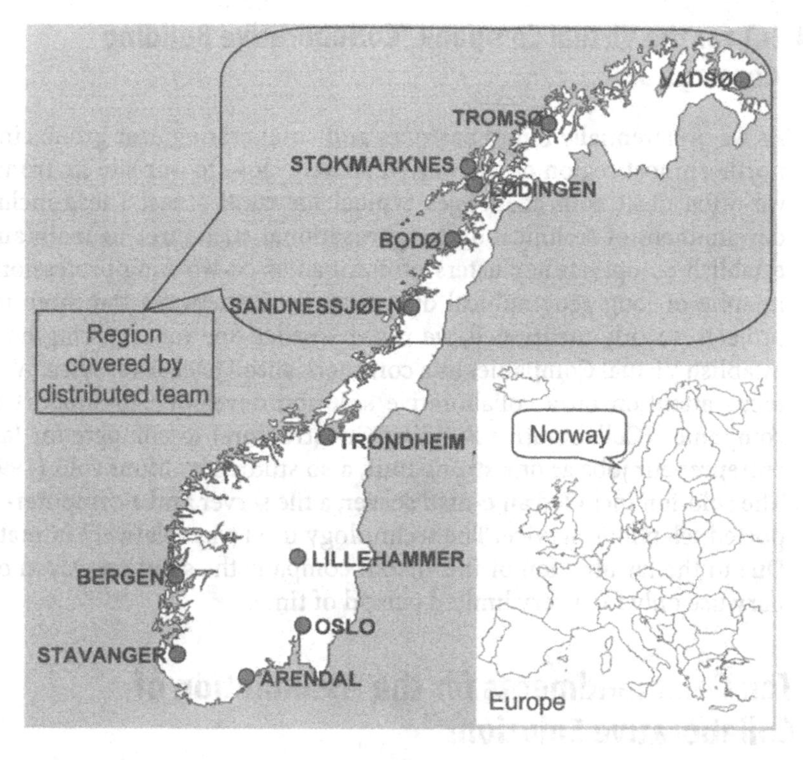

Figure 9.4 Map showing the location of the Health, Safety and Environment teams (Evjemo et al., 1999).

within a co-located team: The numerous "snapshots" (the very presence of your colleagues, glimpses of visitors, discussions at lunchtime, half-written manuscripts at the printer, drawings on whiteboards) give you brief information about what is going on and, to some extent, the agenda of your colleagues. Next, the team members access the same sources of information, and they are close to each other (in several meanings of the word) and thereby develop particular manners of addressing colleagues and their competence and adjust to one another in a fine and not easily identifiable way.

Hence, we built a Lotus Notes application, accessible from the web, with the following features: a view of any event related to customers' requests (updated by the team members), a bulletin board, meeting agenda, appointment list, and discussions/experience exchange. The field study went on for nearly two years, and the solution developed was after the project period adapted and transferred to the other six teams within the department of Health, Safety and Environment. Figure 9.4 shows the location of the distributed team.

9.2.4 ICT for the Virtual Company "Collaborative Building Constructors"

As we preferentially found partners and co-operating user groups in the northernmost region of Norway (relatively close to our site in Tromsø), we often dealt with challenges typical for rural areas. These included development of technical and organisational structures to maintain or establish co-operating clusters of companies co-working professionally in spite of long geographical distances. So, in between the three main projects already mentioned, we ran a smaller one investigating how to establish Virtual Companies as a commercialised telecom service. We will focus attention on a collaborative solution developed to support four companies (Collaborative Building Constructors) to compete for larger entrepreneur jobs as one strong unit, also studied by Munkvold (1998a). The solution included an e-mail server, a file server and a computer-supported telephone service. The technology used was relatively immature. Due to the termination of the virtual company the solution was in ordinary use only for a very limited period of time.

9.3 Ten Commandments for the Assimilation of Collaborative Solutions

Based upon our experiences from the projects presented in the previous section we have developed a set of commandments that we apply as a

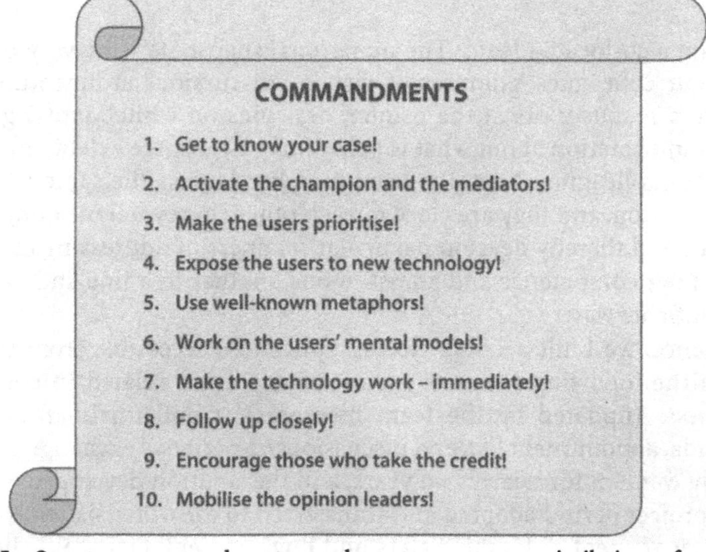

COMMANDMENTS
1. Get to know your case!
2. Activate the champion and the mediators!
3. Make the users prioritise!
4. Expose the users to new technology!
5. Use well-known metaphors!
6. Work on the users' mental models!
7. Make the technology work – immediately!
8. Follow up closely!
9. Encourage those who take the credit!
10. Mobilise the opinion leaders!

Figure 9.5 Our ten commandments on how to support assimilation of collaborative solutions.

basis for our further work related to the assimilation of sector-specific collaborative solutions (see Figure 9.5). In the following we present these commandments, together with illustrative experiences from the projects.

9.3.1 Get to Know Your Case!

This first commandment is reflected in our general approach to introduce collaborative solutions. In order to develop tools that work well we need to "get under the skin" of the users, know the "rules" of the organisation and also understand the sector or wider surroundings that the organisation is part of. Such an understanding can be obtained by studies of formal documents (organisational charts, procedure checklists and so on), charting of workflow and on-site observation of certain operations. These relatively unobtrusive techniques need to be accompanied with more interactive actions inherent to the PAR approach described in the introduction of this chapter.

A thorough understanding of the organisational context might for instance uncover differing perspectives of cost/benefit within the group involved. To be aware of and, even more important, be able to meet this challenging situation from the very beginning, will pay back in renewed willingness to cope with extra working loads in the weeks or months to come. Frustrated users might turn into satisfied users simply by the fact that their situation is understood, or in certain circumstances, that some kind of compensation is given. This point is illustrated below by one example from the telemedicine project and one example from the local politicians project.

The Teleradiology solution was developed to enable the radiologists at the University Hospital to diagnose X-ray images at a small regional hospital from their own office. They had previously travelled 300 km once a week to do this job. The potential savings were high, but the expected enthusiasm from the radiologists did not appear. On closer examination it appeared that despite the variable driving conditions, the two-hour journey to the small regional hospital was seen as a relaxing break in an otherwise stressed day. In addition there was no compensation arrangement in place to cover for the new work they had to do, and they also lost their mileage compensation and food allowances. When they were compensated correspondingly, a serious obstacle to the solution was removed, and the radiologists gradually changed their attitude in favour of the solution.

This story illustrates that it sometimes is difficult to uncover the real barriers to adoption. The radiologists themselves were probably not particularly certain about why they missed the "good old days". We should therefore be prepared for the fact that users are neither conscious, nor perhaps wish to reveal the real reason for the lack of enthusiasm in their attitude towards a new solution. In the close surroundings of the local politicians we realised in the early phase that the local government officers became frustrated. It was still hard to find a solution, but we believe that the positive dialogue we managed to establish cheered them up and made them go on ("for a better tomorrow").

Through the Local Politicians' Channel the local government officers were told to distribute electronically the information that should be handed to the politicians. This change of procedures had the potential to save both time and money because of reduced copying and distribution costs. The politicians appreciated this, because they got the information earlier than before. But, due to the fact that a lot of attachments were not in electronic form (no suitable scanning service existed at that time), the manual and electronic distribution of issue information needed to exist in parallel, which naturally posed extra (not less) work on the officers, at least for a period of time.

Other researchers have highlighted the cost/benefit problem, not least in the early phase of the electronic group calendars (Grudin, 1988, 1994b; Bullen and Bennet, 1990). Those who most often call a meeting, e.g. managers, usually benefit, while the rest of the group is given a little extra work in keeping their electronic calendars up-to-date. In that way they can find timeslots when everyone in the group can be present. For many group members this means double entry, in the paper calendars and in the electronic one. It is therefore not without reason that the electronic group calendars some years ago were called "one of the most available but least useful co-operative technologies" (Grudin, 1994b).

The group calendar system is now gaining in popularity. This might be explained by individuals finding compensation for the eventual extra work that follows with entry of electronic calendars. The calendar systems function to a greater degree as activity support for individuals as well, with personal activity lists, handy address lists, and the possibility of connecting the address list to e-mail, telefax and telephone. This example shows that not only a change in procedures can change the cost–benefit balance; technological developments (as in the case of electronic calendars and hand-held devices) can change this balance as well.

Collaboration technology launches new styles for collaboration with more opportunities, which are not welcomed by everyone. There might be "substituting" or "straw men" arguments that complicate the assimilation and, unless they are uncovered, make it difficult to explain the identified effects (or the lack of such). In the case of local politicians the ICT solution introduced a possibility for previously peripheral members to come closer to the decision-making processes in the group. Other

group members, however, were not very enthusiastic about this development. Knowledge is power, and when information suddenly is in free circulation, those who previously based their position on being information distributors to the group felt threatened. For power wielders it can be uncomfortable to realise that information takes new "un-controllable" routes, but at the same time there are few who will intimate that this is the reason for their lukewarm interest in the new technology. It is easier to blame technical problems, poor design, etc.

9.3.2 Activate the Champion and the Mediators!

A champion within the user group is crucial to successful projects. He or she is the one who fancies the idea immediately: "I like this idea. Let's go for it!" He is also the one with personal qualifications to front the idea. The champion does not necessarily belong to the management, but he certainly needs to have some formal or informal position in the organisation that makes it possible for him to initiate change processes. He clears away hindrances of both practical and bureaucratic nature, and ensures good working conditions for everybody involved. We experienced how important this person might be, in our implementation of the collaborative solution for the local politicians:

In the implementation of the "Local Politicians' Channel", the 33 politicians of a rural municipality were provided with PCs and communication lines and a collaborative solution specially made for them.

The mayor was very soon identified as a champion. In addition to personal enthusiasm and understanding of the potential benefits of the solution – even beyond the project's limited aims – he had a rather unique and strong position in terms of political support. He spoke with great authority and actively promoted the project through several channels. Within some few weeks we had the council's formalised commitment to join the project and pilot the technology we were about to develop. We know we could not have come to this stage without his willingness and belief in the idea of the project.

Unfortunately he died six months after the solution was introduced. His successor as a mayor inherited the formal responsibility for the adaptation process. He did not seem particularly interested in the solution, but still promised to follow up the deceased mayor's good intentions. The personal spirit and engagement was, however, gone with the past mayor. The council members continued to use the system, but they were more likely to give up (for instance in cases of technical problems) and return to the old way of working. There was no one there to catch them. When the mayor did not react upon this kind of omissions, nobody else had the power to do so.

This loss of a champion was obvious throughout the project period (Watson et al., 1999). We looked for another person in the council to fit the champion role, but did not succeed. Several of the council members fancied the idea and found the solution most useful, but they neither possessed the necessary position in the group to advocate it nor the personal qualifications needed.

In some of the political parties, however, we observed the presence of technology-use mediators. This was reflected in a more extensive use of the system among a party's members and also in more innovative uses of the system, for example in the co-authoring of position papers for the budget discussion. Through the part-time employment of one of the most experienced users for doing support, we recognised and rewarded this mediator role which complemented the champion role in the first phase of the project and next became the most valuable contact between the user group and us.

So, like the champion, potential mediators should be identified as early as possible. These are leading users, who understand the tool's possibilities (and limitations) in relation to actual work tasks and the further user settings (see also Section 9.3.6). The mediators are involved in the convention rules and patterns of use, which gradually develop. It is therefore decisive that they are followed up closely. Okamura et al. (1994) emphasise the mediator role, and allege, "Mediation – as an ongoing and organisationally sanctioned intervention – may be particularly effective at overcoming some of the problems in CSCW use identified by other researchers". In the politician field study we saw that the patterns of use varied greatly between party groups, and we explained this phenomenon by the existence of mediators (Ytterstad et al., 1996). If the mediator loses inspiration or faith in the solution it is difficult to convince the rest of the group. It is not least important that the mediator appears to involve the other group members.

9.3.3 Make the Users Prioritise!

The collaborative solutions used in our projects are based on a mixture of identified user needs and qualified guessing – with the developers in charge of both approaches. Most often the list of wanted functionality goes beyond the resources available at the time, either of economical, technical, or time-consuming reasons. Particular organisational and security circumstances might also impact the final decisions on what features to be implemented. Anyway, there will be an indisputable need of prioritising and we strongly recommend that the users participate in this process. Our efforts working with the Health, Safety and Environmental team illustrate this situation.

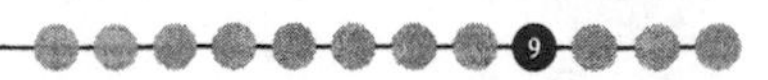

The HSE Net was developed to make it easier for the distributed Health, Safety and Environment team to share information, collaborate on specific tasks and get more involved in each other's tasks and clients and strengthen the feeling of group belonging.

The initial period was characterised by repeated discussions on what the solution should look like. After several meetings with the users we summed up the discussions in a project directive – written as a form of contract – where the main idea was formulated and the first draft of the solution was presented. This document released a lively discussion where both misunderstandings and disagreements were uncovered. We had to go through the initial ideas, and in particular the technical limitations that happened to be more restrictive than we (and the user group) had expected. We ended up with a revised version of the design; much more detailed than the previous one and committing for both users and developers.

Making the users prioritise means they are "forced" to go through their most basic needs informed by the actual limitations. Perhaps they do not agree on what those basic needs are. Or, the leader has other needs than the rest of the team, and these particular needs might be in conflict with other essential needs. In other words we should be particular about involving a sufficiently wide spectrum of user roles. Furthermore, as researchers, we might have specific hypotheses to be tested, which means that some system features are regarded as a must. Hopefully, there are no big deviations here.

With these problems in mind, Yourdon (1996) has devised the following two rules-of-thumb: (1) find out what is good enough; (2) divide the development timetable into several phases/portions (particularly relevant for larger systems). The critical point is of course to decide what is "good enough". Yourdon postulates three questions to help us with the first rule, and the users should answer them all:

1. Which needs are absolutely *essential*? The solution would be unusable without covering these needs.
2. Which needs are *important*? If these needs are not met, the solution would be difficult or long-winded to use.
3. Which needs can be called *accessories*? If included, these will make the solution more interesting or enjoyable to use.

Even if the questions are simple, they are not necessarily easy to answer. Anyway, they will initiate discussions and produce some kind of quality test on the decisions done so far. The project as a whole will benefit from this in several ways. First and foremost conflicting interests among the users might be uncovered at an early stage. This might explain unexpected frustration and negative attitudes and perhaps lead to revisions of the solution (see also Section 9.3.1). Next, this discussion might establish a common understanding of success criteria, which are important as guiding lines through the adaptation process but also in a final evaluation process. What did we actually agree on as the most important issues? This is of course addressed by the formal project goals, but might still not be fully understood or accepted by every member of the team.

If we return to the Health, Safety and Environmental team project, we took this process seriously, but still not seriously enough. There were obviously divergent opinions in the team upon the purpose of the project. The leader had specific reporting needs, which we noticed but did not understand as fundamental to the collaborative tasks. As a matter of fact, they were not, but still turned out to be decisive for the leader's interest in central parts (from our point of view) of the solution. He was enthusiastic about the project, but our user logs uncovered that he hardly used it himself. The attitude of the leader is important, both as an example to the rest of the group ("he uses it that way – we have to follow up"), but also because he is the one initiating tasks and directing new routines within the group. At a later stage of the project we finally implemented the reporting features and we noticed positive changes on his usage frequency.

Yourdon's (1996) second rule-of-thumb is more straightforward and the benefits easier to see: by dividing the solution into modules or deliveries you are able to prototype the solution and also provide the users very quickly with a first version. Through practice their engagement is maintained and the further discussions on functionality are made easier. It is also easier to identify potential sources of error and blind alleys when you expose the system to the users early. By giving the users insight into the developing process – and taking account of their comments – the crucial feeling of ownership to the solution is established. However, you also risk losing their interest if the system fails, which is likely to happen when you put early versions into production. Here we must distinguish between enthusiasts, who experience error situations as a creative brain stimulus, and the rest, normally representing the majority, who react to system failures with alarm and frustration (see also Section 9.3.7). Lastly, if the manager is reluctant to make up his mind but has to commit himself to the project in the early stage, then he is likely to feel more confident if he can bail out at upcoming predefined milestones.

9.3.4 Expose the Users to New Technology!

If you ask a user whether he needs new technology or new features in the existing one, he would often give you an answer like "No. Everything works fine!" You may get this answer because the user has loyally adapted his way of working to the requirement of the system or he does not know the potential of the technology in his actual work. In general, end users often do not have insight into the potential of the technology and cannot always predict their own needs (Prinz et al., 1998). Working together with them for a while, you might also uncover that they do not know their colleagues' routines and are not able to recognise how their routines interfere with those of their colleagues. On the other side, as system designers we are not able to achieve complete knowledge of the working processes within the group and certainly not all the group dynamics and individual preferences. To come up with a valuable solution the users and the designers have to educate each other. This might, for example, be done through exposing the users to new technology.

Turning to the Health, Safety and Environment team again, we can see how this strategy influenced on the adaptation of the solution.

The Health, Safety and Environment team "met" all together only once a week in an audio conference. Otherwise they called each other on a regular basis or whenever needed, but this half an hour was their opportunity to share formal and informal information in plenum. We decided to implement a bulletin board into the HSE Net (see Figure 9.6) without asking the users in advance and we did not give any suggestions about how it could be used.

By observing the actual use over several months it showed two interesting facts. The bulletin board became an informal (virtual) meeting place for the group and it revealed a need for a group calendar. The latter led to the implementation of a simple group calendar. The group was not aware of this possibility during the initial design phase of the application.

Actually we here chose to provide functionality that in the initial phase – if suggested – could have been categorised as an accessory (see Section 9.3.3) and consequently not been implemented. As the implementation of the bulletin board was rather uncomplicated (not very time consuming to implement and easy to remove), we found it well worth an experiment.

The designers should keep in mind the technological potential when collecting information on the work processes to be implemented. By implementing promising features in the actual collaborative solution, the users are exposed to technology that may lead to user-driven suggestions on new or changed functionality in the system. The designers should listen for such suggestions both before the initial version of the system is developed and in meetings with the users after the system has been put in use. Listening carefully you could as well come along with ideas for

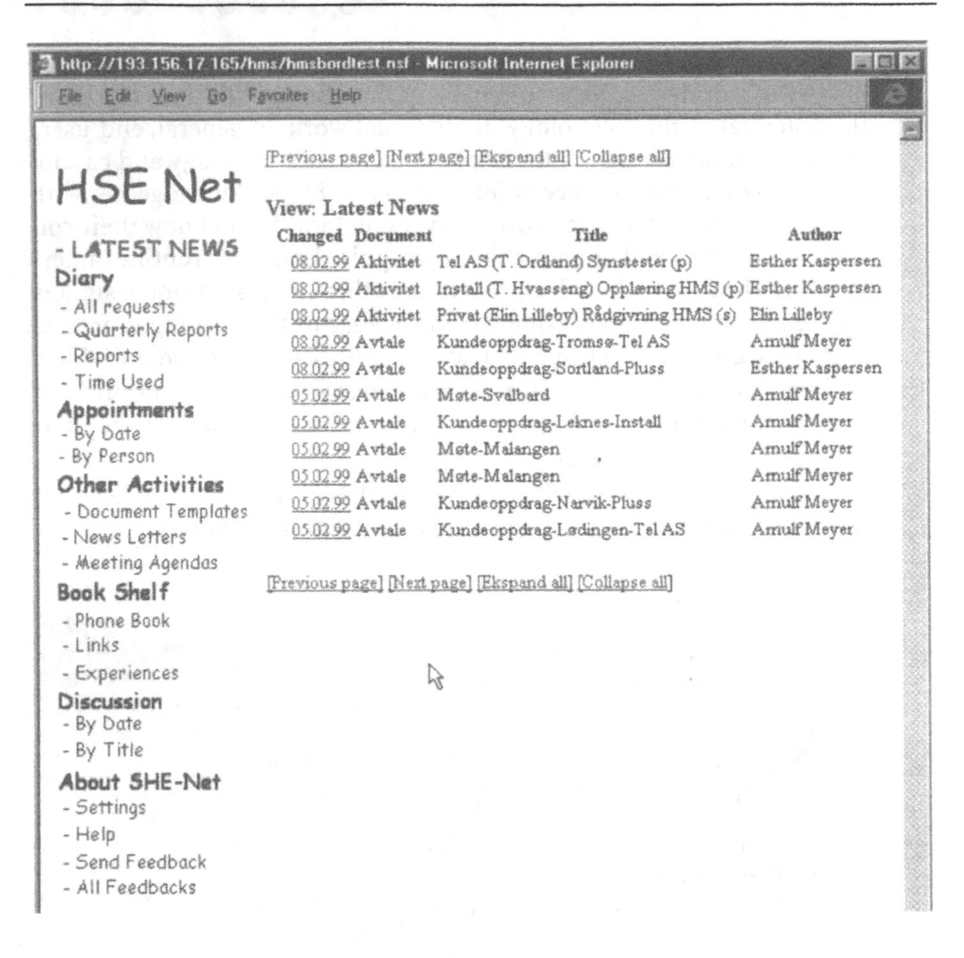

Figure 9.6 Screen dump from the welcome page of the HSE Net application.

additional products or new project plans – as in the following suggestion proclaimed by one of the local politicians:

> I also think there are several ways this project could be taken further both in terms of user groups and applications. The Local Politicians' Channel would be perfect for connecting local politicians with politicians in other municipalities or on the regional or national level. Combined with the Internet, this would give access to a whole range of applications and information sources relevant to the politician.

9.3.5 Use Well-known Metaphors!

When you recognise your own statements in a joint textual delivery or you find your company's logo among the advertisements or printed on top of the website, you experience some kind of recognition and your attention is caught. Sitting in front of the screen you prefer your own settings to occur; you have selected the colours of the background, the fonts and the size of the windows. You feel comfortable within this well-known environment. Big companies often put additional efforts into adjusting commercial software packages so that these better fit the company's organisational structure, its particular working routines and document flow, etc. These are often cosmetic changes, but still important as they might bring forward some feeling of confidence, ownership and understanding. Well-known metaphors and use of internal jargon might even make you think: "This application is tailored for us and has consequently taken up our needs".

When developing the collaborative system for local politicians, we directed considerable efforts into the design of the user interface (Ytterstad et al., 1996). We created graphical directories to emphasise the organisation chart, and the different chairmen and committees were easily recognisable as well. We had further scanned in the political parties' emblems as obvious navigation signs (see Figure 9.7). By using a graphical directory, the system presents symbols, pictures, or other graphical information readily recognisable to users. This interface design was successful not only because of the system's ease of use but also due to its ability to create an illusion of proximity, because of the comforting sight of familiar symbols. The use of party logos as pointers to directory information seems to have been a crucial design decision. This tailoring undoubtedly contributed to the feeling of ownership, both to the solution and the project.

The local politicians constitute a group of users with great diversity, both in background and computer experience. Some of them had never used a keyboard before and spent considerable time just finding the spacebar! Surprisingly, these potential problems were quite easily handled. Users with extremely limited prior computer experience were able to perform the basic operations needed for communicating. The users identified the system's intuitive interface and low user threshold as reasons why they were able to learn easily how to use the system.

In the adaptation of the HSE Net the users asked for particular report headings and well-known guidelines to occur on the screen. These were obviously important to make the users adopt the system. In the local politicians' case we made use of this strategy from the very beginning.

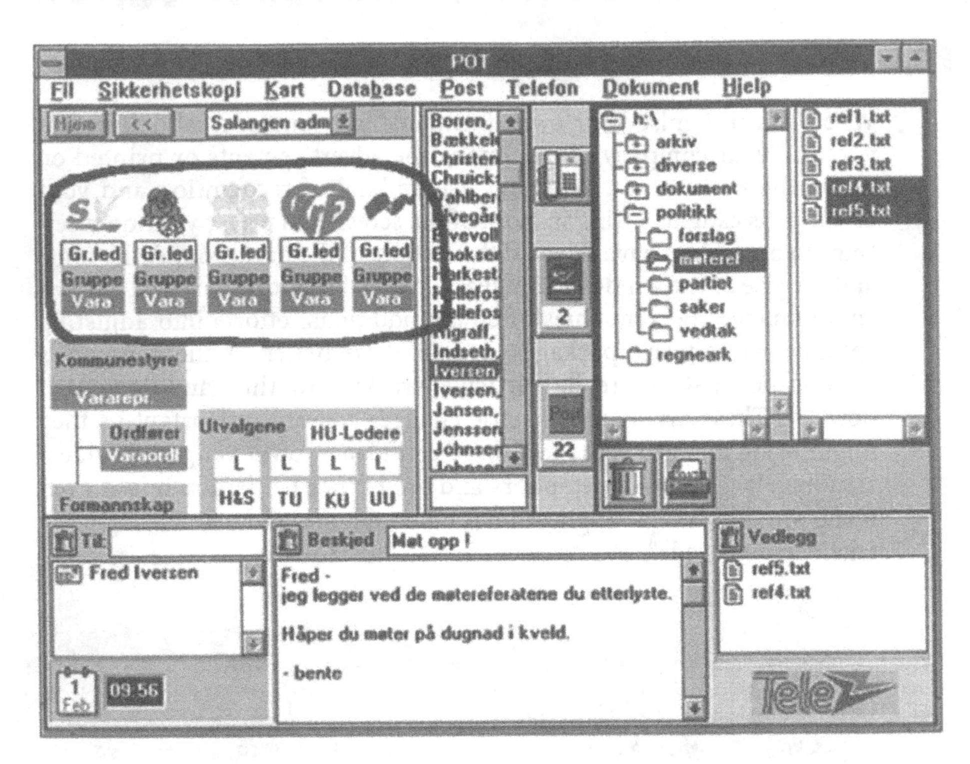

Figure 9.7 Graphical interface including organisation chart and party emblems as part of the collaborative solution made for local politicians.

9.3.6 Work on the Users' Mental Models!

As a user is exposed to new technology and possibilities, it is important to focus on his tasks and roles within a group, and how his ways of working also influence other group members. A collaborative tool should not be introduced in isolation – the benefits arise when the individuals use the technology with an eye to the group as a whole. This is why it is important to:

- teach collaboration technology in groups;
- establish common conventions on how to use the technology.

At the very start of the local politicians' project we completed a joint training session for the whole council. The district arts centre was hired for the occasion, and was, just for that day, the most computer centric arts centre in the whole country. A trailer had delivered 33 PCs the evening before, and the same number of modems and telephones, as well as an exchange in the courtyard. The machines were placed in a horseshoe formation – as at a banquet – with the cables in the middle, where the exchange was enthroned as a symbol of community and fellowship (see Figure 9.8). The participants were mainly without special knowledge of computer technology, and we had to depict, clearly, neatly and physically, how the individual machines worked and how they could interact with the telephones.

A repeat course was held some months later. In addition, individual party group training sessions for establishing internal conventions on how to use the technology, were held. We identified the most well functioned groups and transferred very gently their ideas to the others. These conventions were very welcomed.

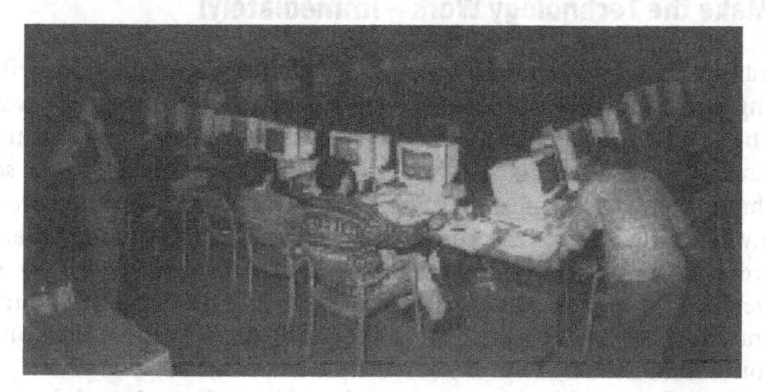

Figure 9.8 Politicians attending a one-day training session before taking their PCs home.

Orlikowski (1992) alleges that collaboration technology is often used as if it was traditional software (e.g. word-processors or spreadsheets). People do not take advantage of collaboration technology's unique possibilities for supporting collaboration and management. Orlikowski further argues that where the spirit of collaboration does not already exist, collaborative technology cannot charm it into existence on its own. In academic terms, we should here also focus on (1) people's mental models connected with technology and their own jobs, and (2) structural characteristics in the organisation (strategies, standards, reward systems, etc.).

Collaboration technology provides the user with the possibility of collaborating independently of time and space. This aspect will not instantly lure users into adopting and exploiting it. Their mental model of computer systems connected to their own work routines must be changed from "office support tools adapted to personal needs" to a "collaborative tool which includes and brings to view colleagues and their activities". Most people need help with that. Whether the above-mentioned telephone exchange in fact contributed to changing the politicians' mental models cannot be asserted with certainty. However, as a pedagogic agent it was useful at the time.

At some stage in the introduction phase the usage pattern sets, and this might very well happen at a non-advanced level. As we visited the party groups in the local politicians' project, we made them explain their working routines in detail and interrupted their presentation with questions like "what triggers the use of the solution in this particular situation?" or "how can the system possibly help you here?". They were in a way forced to argue for their choices and preferences, which finally brought forward alternative ways of working. This undoubtedly stimulated the establishment of new mental models.

9.3.7 Make the Technology Work – Immediately!

Introducing collaborative solutions, often in combination with re-engineering of work processes, puts a certain pressure on the workers. They neither have extra time nor patience to deal with technical problems. The solution must work immediately; you never get a second chance to make a good first impression! And it must work for every one involved, not only for the pilot users or the technicians among them who are able to troubleshoot and fix the small problems as they occur. Detailed testing procedures should of course be developed, and further – and most important – the testing should be done by more than one person simultaneously, to provoke multiple user situations.

We will never forget how technical problems demotivated the users of the promising solution for the Collaborative Building Constructors:

> We developed a virtual private network to support six collaborating entrepreneurs in the northern parts of Norway. Due to various technical problems and unstable ISDN lines, four months elapsed between the first and last entrepreneur could access the network. This led to a situation where workers were taught how to use the solution several months before they finally could make use of their knowledge. Naturally, we had a hard job to motivate them for another try.

Most teams and organisations we have been in touch with have a tight working agenda. Such working conditions make them prioritise their tasks strongly. Only the most crucial and business-critical tasks tend to be carried out, with less important ones being put on hold. Even though these partners have acknowledged the need for collaborative solutions in their own working processes, finding time to take part in designing, and later on learning how to use the technology, has been difficult. We have also experienced leaders in a team who have eagerly decided to start using an application, but turned angry when they faced the fact that they had to spend time learning to use the solution or because their PCs started "acting strange".

A malfunctioning system may turn users away because they may be unable to get their work done satisfactorily. Users who feel uncomfortable in using the system for other reasons, get yet another argument for not using it. Whatever the reason, if workers stop using a collaborative solution it affects the other users because they do not get the full benefit from the system. Remaining users have fewer peers to share the system with, which again may lead to new dropouts.

A well-functioning system creates good ambassadors as opposed to saboteurs. Saboteurs are users who do not like the system and express their bad experiences to other users in the organisation. Saboteurs may also tell other potential user groups about their lack of satisfaction with the system. Such workers may bring a collaborative solution out of business even if it is promising for its intended users.

Even if a malfunctioning system is not generating dropouts, it may lead to less user involvement. Designers of this kind of systems rely heavily on the users' goodwill and commitment to get them to describe their working processes, and later on how well the system fits their work routines.

9.3.8 Follow Up Closely!

During the introduction of a system the users typically report errors and questions on how to use the system. To minimise the possibility of a user getting stuck, immediate support is necessary. Especially in cases where a collaborative solution is replacing an existing system or manual working routines, there is a need for quick responses to user problems. Otherwise the user may fall back on earlier working habits and his confidence in the new system may decrease. Within the Telemedicine Project we found it necessary to be co-located with the users for some time to take care of technical problems as they occurred. In the Health, Safety and Environment case we tried to follow up the users very closely, not only in the first critical phase but also throughout the whole project period.

Collaborative solutions reach their highest potential when the whole team is closely engaged. Following up all the team members on both technical and organisational issues is therefore particularly important. In

The distributed Health, Safety and Environment team was told to call us for help anytime within their regular working hours, or if preferred, send us e-mail. At least one person in the support group was designated to answer questions from the users, and make sure that the system was accessible. We experienced users that seemed positively surprised, and due to the short time of response they seemed to consider us as being geographically close to them (even if we were not).

We also implemented a feedback mechanism in the HSE Net solution where the users could report on errors. Error messages were accompanied with contextual information, which was valuable for us in the process of correcting the error. This feedback mechanism was also used to mediate suggestions and wishes for system changes and new functionality. It was important that we checked for new requests every day and reported in detail on the actions done to meet the request (Received, In progress, Finished). Input concerning greater design changes was put on the agenda of design meetings.

contrast, a single-user solution like a word processor may be used independently of other users' actions. As discussed in Section 9.3.7 dropouts may occur if the collaborative solution fails technically or if the individual user does not see the benefits. Obviously, workers who do not use the solution leads to less information shared through the system, which has consequences for the remaining users.

Meetings concerning the way the system fits the actual work process must be held on a regular basis. This will uncover the different ways of using the system, and most important, it may reveal why users are *not* using it. Congruent use is required to make the system work optimal within the group (see Section 9.3.6). Catching *why* some workers stop using the system might lead to design changes. Further, discussions within the work group may also lead to suggestions on how to activate non-users.

9.3.9 Encourage Those Who Take the Credit!

When the actual users of collaborative arrangements talk positively of these, this often has a great impact on other potential users. The user then acts as a kind of witness or ambassador for the solution, and other users become interested in trying out something similar. In order to facilitate such an effect we propose supporting the persons who take credit for a solution (even if they might not be the persons having spent most time on making the solution work). Providing help in writing presentations, co-authoring scientific papers, setting up demonstrations, and pointing to the user as an innovator in press releases may help in accomplishing this effect.

In the Telemedicine project we had (mostly) joint interests with the involved healthcare personnel. We deliberately supported them in telling the story in papers, at conferences and elsewhere. The effect in terms of ownership to the solution increased. Further, the healthcare personnel became good ambassadors for the solutions. As a matter of fact we followed this commandment to the extreme in the establishment of a dedicated telemedicine centre:

> In 1988 Telenor embarked on the Telemedicine project in close cooperation with the University Hospital of Tromsø. After a few years of successful trials demonstrating the potential, the University Hospital established a separate department with specific responsibility for telemedicine. For a period of 18 months Telenor "leased" five researchers for free to build this new department together with University Hospital personnel. During this period the healthcare personnel built up their own competence and became the telemedicine experts for the future. They gained international recognition and got the responsibility for further telemedicine research and diffusion at a national level through the establishment of the Norwegian Centre for Telemedicine.

To encourage those who take the credit does not necessarily mean that the wrong man or wrong institution is given honour. Within our projects the healthcare personnel co-working with us were crucial to the gained results, and when the time was ripe to "tell the world" about our experiences, these people were the right ones for doing that. One reason is simply that healthcare personnel listen to other healthcare personnel. They can tell the story from their point of view and add confidence to the trials with new technology. In these circumstances we act as the agents of change, leaving the stage for the next crew to take the credit and bring the work further.

9.3.10 Mobilise Opinion Leaders!

Compared to other ICT solutions, collaborative solutions seem to need more support and more time to be fully adopted within an organisation. To meet this challenge, we believe in bringing forward successful stories, and furthermore, to mobilise opinion leaders.

Opinion leaders might also be referred to as a third party (the first party being users and the second being ICT providers of various kinds, telecommunication operators included). To fulfil the role as an opinion leader the worker union – or whichever interest organisation being addressed – must possess idealistic attitudes and act as a neutral part with no economical benefit from the promoted products.

While the Local Politicians' Channel was explored and used by the selected council, we contacted central members of the Norwegian Council's union to make them aware of the on-going experiment. They were partly informed by the mayor of the council, a dialogue that seemed to have vanished with the mayor. Serving the interests of the 430 Norwegian local councils, we regarded this organisation as the most suitable opinion leader for this particular solution. With one successful implementation, and the union speaking enthusiastically about it, the take up of the solution within other councils would have been likely. The union did find the concept interesting, but when the council chose to assess the benefit of the solution from an economical point of view only (Watson et al., 1999), we unfortunately missed the success story in the dialogue with the opinion leader. Also technology development and the wider availability of the Internet had made other solutions more viable by that time.

The Norwegian Centre for Telemedicine, Norwegian Medical Association and Norwegian Patients Association are together with political bodies obvious opinion leaders for health and medical-related ICT solutions. The healthcare sector will listen carefully to advice from those organisations because they are non-profit organisations and possess their own experts capable of evaluating the true benefits of the solution. Similar organisations will serve as opinion leaders for other sector-specific collaborative solutions.

9.4 Reflections

Through years of field studies we have generated certain guidelines on which we now rely when new projects are initiated. The ten commandments given here summarise these experiences. We have used both failure and success stories to illustrate how we came to this understanding, and we have done so because in the perspective of gaining new knowledge, failures are as valuable as successes. The crucial point is our ability to reflect on *why* we missed and next, our willingness to act differently in the future.

9.4.1 Ownership

Several of the lessons are related to each other in the way they impact on the users and the adaptation process. The feeling of ownership is the most prominent and most wanted impact of this kind – and seems to us to be a precondition to successful assimilation of a collaborative solution.

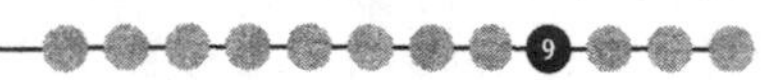

The feeling of ownership is somehow related to trust, confidence and control. Ownership means you are able to influence possible future modifications of the solution, and makes the user feel on top of the situation. He participates in the design and has an image of how the solution is going to impact on the team. We have obtained a positive circle as these positive attitudes makes the users involve even more in the adaptation process, which again strengthens the feeling of ownership – and so on. The owners of the solution love to tell their "story" about it. And this type of external marketing turns the users into the solution's greatest ambassadors.

9.4.2 Long-term Commitment

Although not written explicitly as a commandment, one may read between the lines that there is no easy way to success. Even if the total project period reaches a considerable number of months, the actual trial period seems to be too short. One of the local politicians summarised the field study like this:

> I have a feeling that the experiment was too short. Since it took more than a year to overcome all the technical difficulties, we were left with only half a year to concentrate fully on the usage of the system. Thus we were not able to establish and try out routines within the party groups to the extent we wanted to (Ytterstad et al., 1996).

Furthermore, all parties involved have to work hard and be goal directed with an inner wish for success. As developers or facilitators we have to enter the arena with a mixture of humility and enthusiasm. It is important to understand – and to a certain degree adapt – the users' jargon and the way they communicate and to respect their professional preferences and attitudes. In particular we have to accept any reluctance to the new solution caused by the essential need of serving demanding customers. Introducing new routines while still updating the old archives or running the old routines as a backup, might provide problems or delays for the customers. Their initial tight agendas are not likely to be less tight due to our arrival. Consequently we have to step back and help them run the two streams of routines as long as it is regarded necessary, and make sure they get to do their business. This understanding and respect is essential to establish the necessary trust and confidence. Note that collaborative solutions might be introduced for a smaller group or pilot groups until the solution is trustworthy to avoid double work for many participants. This approach has to be balanced towards the need of involving the whole group to ensure that the core collaborative issues are addressed, even in the initial phases.

On the other hand, a successful assimilation process depends on commitment from the team involved. Users should be prepared to spend some hours in initial meetings, for internal discussions, and idea generation. They should also be aware of the efforts and time needed to establish new working routines and common conventions within the team. The main concern from the users' perspective is to serve their clients or meet the working day's various requests in parallel to the introduction of new working routines. In days of heavy workloads this might be difficult, which consequently means we should choose the introduction period carefully.

As this process might occupy several working hours – which indeed represents costs to the company – it is necessary to involve the management and get their commitment. To be sure that the solution will be fully integrated into the organisation and also developed to fully fit the individual and organisational needs, a long-term commitment is desirable. This is of particular importance in the case of collaborative solutions where the expected effects are not necessarily easy to identify. The results do not occur during the first fortnight, at least not the more subtle ones – related to group dynamics, group maintenance and individual satisfaction. So, a management that is willing to pay the costs (and thereby shows that it understands the potential of the solution) is more likely to keep trying and fighting for good results than the one giving you a chance as long as the solution does not provoke negative feelings among the employees or there are unexpected costs coming up.

We also want to emphasise the necessity of close co-operation between developers (or facilitators) and the users. In this context the user is the one who understands the possibilities and limitations of the team and the company, and as technicians *we* are expected to provide similar knowledge on the technology. Being – most often – the initiators, we are responsible for merging these knowledge pools. This means we need sector-specific knowledge, as well as organisational and sociological knowledge within our department. So far we have succeeded in including a mixture of competence within our project groups.

9.4.3 On the Importance of Technology

Our technical background puts us in the position to assist the users and clear up minor technical problems, also outside the scope of the project. This knowledge is welcomed and brings confidence into the setting. Furthermore, we have brought with us "gadgets and gizmos" and without any particular purpose in mind, exposed the users to advanced or emerging technology. In this process of sharing knowledge the potential of technology is visualised, which might bring forward ideas for additional or changed functionality into their own solution, and also mediate to the users what kind of skills we might contribute to the project. These background activities bring confidence and trust to the relationship between users and developers.

Technology is important! And there is no way to succeed without a solution that works – each time, every day. We have emphasised that this is crucial during the first introduction phase, but technical failures will always be a threat to successful implementation – because of the costs to renew trust and confidence. However, several of our commandments focus on non-technical issues and illustrate the importance of well-organised adaptation processes and agreement upon user conventions, the need for changed mental models, the value of enthusiastic champions, devoted technical support and clever mediators within the team.

As shown by the stories presented, the assimilation processes do not always follow a pre-planned schedule. And this is not necessarily a disaster. It would appear that the ability to see new possibilities in chaos rather than a nit-picking search for a recognisable pattern gives the best results. Orlikowski (1996b) illustrates this point when she says that it is often the unplanned results that are important. It can be dealt with by looking at the unpredictable changes, and see them in connection with the technology implementation, take them seriously and meet them graciously. This means that one must be prepared to change direction while implementing, and accept that the situation is constantly changing. Change is based on the participants' own actions, their adaptability and improvisation capability in relation to the possibilities (and hindrances and disadvantages) that new technology represents. This perspective occurred as a reaction towards the more traditional evaluation methods, where one anticipates changes as foreseeable and controlled, and where the technology itself controls the development. Change occurs over time and change is a continuing process (ibid.).

DeSanctis and Poole (1994) describe in their Adaptive Structuration Theory a dynamic perspective on how to justify effects due to technological solutions, considering the effects of information technology to be more a result of how people choose to use it than of the technology itself. This also explains how effects (and the way technological solutions happen to be used) in many cases diverge from the original intentions. Studies have shown that identical equipment and software applied in similar organisations may have different effects (Kraut et al., 1989), and our understanding of the change process (when driven by technology) is therefore concluded to be incomplete. Kraut and colleagues suggest a model with broader compass, where work routines, company strategies, physical space differences, and production and quality specifications are emphasised as "technological elements" in the same way as the technology itself. Consequently, new technological designs cannot achieve maximum results on their own because they are separate from the human component with which they have to interact.

9.4.4 Sector-specific Solutions

So far we have conducted field studies and gained detailed knowledge on several sectors. These initiatives have given insight into different business areas, as well as an awareness of sector-specific difficulties related to successful implementation of collaborative solutions. The sectors differ in so many ways: overall goals and willingness to see alternative ways to reach them, organisational structure, economic situation, technological competence and interest, degree of autonomy and so on. These issues have become decisive for the way to act.

For instance, in working within the healthcare sector we saw more easily how to reuse solutions or parts of a solution and also how to build on the experiences already gained on organisational and personnel related challenges. Even though the disciplines in the healthcare sector differ, the personnel share a simple and assembling mission: to cure human beings. Furthermore, the personnel are trained within the same educational system, and thus share a common understanding of their roles and practice. This means that the analogies, both regarding challenges and possible collaborative solutions, were obvious to catch for developers as well as for users.

The project "Telecommunications within Local Public Sector" was initiated and implemented in the same way as the Telemedicine project. But here we faced a sector built up by separated units, with nothing but their employer in common. How could educational and cleansing personnel share perceptions and ideas on how to collaborate? How could we find the core or the grounded spirit within this very dispersed sector? As time went by we concentrated on smaller groups and their particular challenges, and did not pay much attention to the whole sector as a unit. This change of strategy emphasises the need of targeting users that actually have something in common.

The small and medium-sized enterprises represented a business-oriented mentality, different again from the two public – and thereby non-profit-oriented – sectors we had studied previously. These companies had to face the simple rules of business, which meant, for instance, limited economic capability (and willingness) to accept failures. Involvement in other sectors (fishery sector, the marine sector and the travel and tourism sector) supported our belief in the necessity of developing sector-specific solutions or at least make sector-specific adaptations.

But is this approach special to collaborative solutions? And what about the challenge of generalisation? We would anticipate this statement to yield for all kinds of ICT solutions, but the great complexity connected to the implementation of collaborative solutions, makes this point even more actual for this particular category. People's everyday life is complicated, and so is their working day, their working routines and their relationships with colleagues. A sector-specific solution must make it easy

for them to recognise how this particular solution might solve their special needs. But of course, we do not suggest starting from scratch whenever you involve new teams. There are general components suitable for reuse, but functionality and design have to be adjusted, and this process calls for user involvement and long-term commitment.

9.5 Summary

In this chapter we have presented our experiences from more than ten years of work in the field of implementing and introducing collaborative solutions. We have chosen to state our lessons learned in "ten commandments" that need to be considered in order to make a successful implementation of collaborative solutions.

The set of commandments focuses both on technical and organisational issues. We argue that technology *and* organisation are non-exclusive parts that have to be taken into consideration when introducing collaborative solutions. Practitioners should be informed that both parts might influence on each other, both in early stages of the development of the solution and later on when the system has been used for some time. A collaborative solution may lead to changing work habits in an organisation. Likewise, user experiences with a collaborative solution may call for changes, or added new functionality.

Acknowledgements

We would like to acknowledge some of our colleagues not getting credit through the list of references. These are: Deede Gammon, Rune Hamnvik, Aslak Kjølstad and Bjørn Harald Pedersen. In particular, we would like to thank our colleague Pål S. Malm who was a co-author on an early draft of this text. Additional information might be found at http://www.tft.tele.no/netto/.

10

Implementation and Use of Collaboration Technology in e-Learning: The Case of a Joint University-Corporate MBA

Robert P. Bostrom, Chris Kadlec and Dominic Thomas

10.1 Introduction

This case study presents the lessons learned from the implementation and use of collaboration technology in the ongoing University of Georgia (UGA) Terry College of Business MBA programme created for the PricewaterhouseCoopers (PwC) North American Consulting Group. Reference to PwC or PwC Consulting will denote this specific group, not the entire PwC organisation. PwC Consulting's primary goals were consultant retention and development through a flexible and customised MBA programme. The Terry College was interested in developing a strong position in the executive education market and an e-Learning infrastructure. The two-year MBA programme created for PwC was launched in October 1998, and graduated its second class in September 2001. The programme has been a huge success for both PwC Consulting and UGA. *US News and World Report* rated the programme one of the top online graduate programmes (Special Report: E-Learning, 2001).

Flexibility in the programme comes primarily from utilising a combination of face-to-face classroom sessions and distance learning. Small virtual learning teams (4–5 people) are used as a critical learning vehicle in the programme. Collaboration technologies are the critical enabling agents for classroom, distance and team learning. The case study presents a model of how collaboration technologies can be successfully used in technology-supported/e-Learning environments. It also presents guidelines, based on the lessons learned, for the implementation of collaboration technology to achieve a successful, "blended" e-Learning programme. The chapter is based on data gathered from the key stakeholders: faculty, students, implementation team, and PwC Consulting and UGA management. Two of the authors were heavily involved in the project: one was overall project manager and faculty member and the other was technical project manager.

The chapter will start with a brief background and introduction to e-Learning and the role of collaboration technologies. This will be followed by an overview of the Terry–PwC Consulting MBA programme and a detailed look at collaboration technologies used within it. The chapter focuses on the time frame from when UGA received the contract to the end of the first year of the programme. The last section will summarise lessons learned from the case discussed as a series of guidelines for those wishing to implement e-Learning programmes. Although these guidelines are derived from an academic setting, they are applicable to both organisational and academic environments.

10.2 Background

> The biggest growth in the Internet, and the area that will prove to be one of the biggest agents of change, will be in on-line training, or e-learning (John Chambers, CEO, Cisco).

> It is about raising the fundamental intellect of the organisation every day. It is what makes organisations win. And inspiring people to learn because the excitement and the energy they get from that learning is so enormous; it is how you energise an organisation. By making it curious, by making it say wow, by finding WOWs all of the time, by creating new learning. That is what making an organisation win is all about (Jack Welch at TechLearn 2001, 29 October 2001).

Many feel that the Internet is perhaps the most transformative technology in history. But for all its power, it is just now being tapped to transform education. At the dawn of the 21st century, the education landscape is changing due to the Internet. The Internet is enabling us to bring learning to students instead of bringing students to learning. It is allowing the creation of learning teams and communities that defy the constraints of distance and time, providing access to learning opportunities that were once difficult to obtain. This is true for the schoolhouse, on the college campus, and in corporate training rooms.

The most common terms used to describe technology-supported learning via the Internet are e-Learning and online learning. We will use the term e-Learning. The power of e-Learning to transform the educational experience is awesome, but it has many potential risks and challenges. We need to develop guidelines to ensure that e-Learning

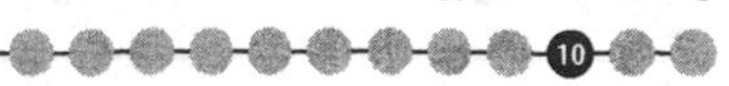

technologies will enhance, and not frustrate, learning. The focus needs to be on learning not technology.

10.3 e-Learning and Collaboration Technologies

The use of e-Learning technologies for the delivery of training continues to grow at an exponential rate. According to International Data Corporation (IDC), US corporate spending on e-Learning will reach $4 billion in 2001, up from $550 million in 1998. By 2004 this is expected to rise to $14.5 billion. We see similar growth patterns in the US for both k–12 (kindergarten to grade 12) and college education. For example, in 2001, over half of all US colleges/universities now offer online courses and over half of all US k–12 teachers now use the Internet in lessons (Johnson, 2001).

During the 1999–2001 time frame, e-Learning has moved from possibility to mainstream – from "will we?" to "how will we?" At the 2001 Comdex Conference, e-Learning was touted the next killer application (Moore and Jones, 2001). Many feel that e-Learning soon will become as ubiquitous as e-mail. The bombings of the World Trade Center in New York City on 11 September 2001 and other events have intensified the focus on e-Learning. Reduced budgets and the inability or unwillingness of people to travel has caused many organisations to start looking at e-Learning and digital collaboration as core mechanisms for supporting and doing business. We see this trend continuing.

Until recently, geography, logistics and scheduling concerns have severely hindered university–company relationships. e-Learning has broken down many of these barriers. Organisations are realising that learning is at the heart of a company. It is the competitive advantage in an organisation. However, organisations have difficulty retaining and developing competent workers especially where a degree is needed to successfully compete for higher positions. Currently, many companies sponsor a valuable employee to return to school; however, they typically lose the services of that employee for the two years of schooling. Companies have little control over scheduling and content in these situations. Many organisations and universities have turned to each other and e-Learning to help solve this problem. This case is an illustration of e-Learning successfully enabling a university–company relationship.

Figure 10.1 shows the time–place matrix, introduced in Chapter 2, used to classify collaboration technology. e-Learning refers to the technology-supported learning activities in any of the four time–place environments. Distance/Distributed/Virtual Learning refers to learning in which the learner and learning resources are separated by space and possibly also time, cells III and IV in Figure 10.1.

Blending became one of the keys to success for the Terry–PwC MBA programme. Almost every aspect of the programme blended different

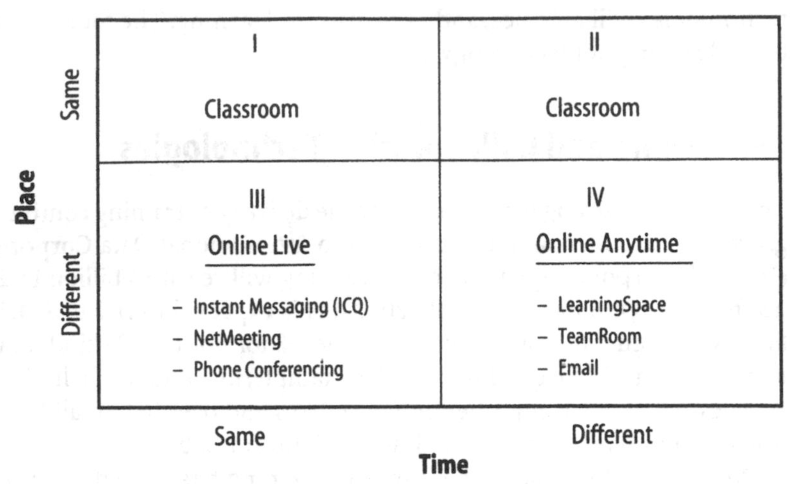

Figure 10.1 Time–place matrix.

approaches. The blend between distance and face-to-face education was a keystone to this programme. A better view of e-Learning technologies and how they can be blended together is captured in Figure 10.2. Figure 10.2 shows three primary learning environments: *classroom, online anytime* and *online live*. These correspond to cells I, III and IV in the time–place matrix. The programme focused on blending these three environments, focusing on classroom and online anytime.

Online anytime technologies support learning anytime, anyplace. They primarily are database-centric creating shared information spaces for learners and faculty to work in at different times and places. Online live technologies provide same time interaction between learners and instructors at a distance through collaboration tools such as chat or a virtual classroom. Online live and online anytime correspond to the terms "synchronous" and "asynchronous shared learning" used in the e-Learning model presented in Chapter 2. Chapter 2 also introduced a third type, referred to as "independent e-Learning" where learners take courses on an individual basis. This type of learning environment would use the same technologies as online anytime, thus we did not include it in our model.

Many view e-Learning only in terms of online live and online anytime technology, both of which facilitate distance or virtual learning. However, the dashed lines in Figure 10.2 indicate that these technologies can also be used to support classroom learning. The Terry–PwC MBA programme uses these e-Learning technologies when students are on campus as well as when they are back on the job. When on campus, these technologies form a "digital surround" for classroom learning. We found that using e-

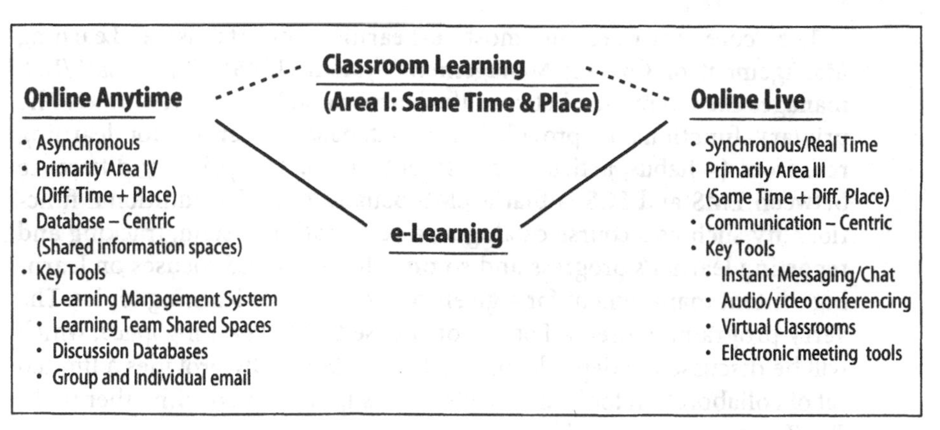

Figure 10.2 Blended learning.

Learning as a "digital surround" enhances classroom learning and provides a great way to introduce students to e-Learning.

Whatever the learning environment, the following technologies are needed to make e-Learning happen:

- *Distribution technology*: technologies that provide information distribution and exchange allowing distance learning to take place. Although the primary focus of most e-Learning is the Internet/Web, sometimes CDs or other distribution technology are used. The focus on the use of the Internet is why the term "online" is frequently used in describing e-Learning.
- *Learning management or content/course management software*: technologies that simulate the experience of a classroom while studying both on-campus and from a distance.
- *Communication and collaboration software* offers a rich, shared, virtual workspace in which instructor and students can interact one-to-one, one-to-many, and many-to-many in order to learn together anytime and anyplace. Examples:
  - *Asynchronous/online anytime tools*: e-mail, discussion databases, streaming audio/video.
  - *Synchronous/online live (real time) tools*: instant messaging, chat, audio/videoconferencing.
- *Course support software* offers a rich set of tools including electronic libraries and other instructional programmes to support specific courses.
- *Server environment software*: technologies to support client–server applications outlined above.

The core software in most e-Learning efforts is a Learning Management or Content Management System (LMS/LCS). A LMS/LCS manages the interaction between the learner and learning resources. The primary functionality provided is a database repository for learning resources (syllabus, articles, assessments, etc.). The primary difference between LMS and LCS is that a LMS usually provides additional functionality such as a course catalogue, a registration system, tracking and reporting learner's progress and so on, whereas a LCS focuses on learning content management for a given course or set of learning topics. The Terry programme used a Lotus Notes based LCS, LearningSpace, which will be discussed in depth later. The LMS/LCS usually provides a limited set of collaboration tools that needs to be supplemented with other tools. The Terry programme did this.

From an e-Learning perspective, collaboration technologies can be viewed as the primary tools that one can use to facilitate learning through collaboration, collaborations between teacher–student and student–student. In Chapter 2, e-Learning technologies were classified in the category for integrated collaboration technology, as these incorporate a variety of digital collaboration functionality: communication (e-mail, instant messaging, streaming audio/video), shared information spaces (LMS/LCS, discussion databases), and coordination (calendaring and scheduling). Although a majority of e-learning technology is collaboration based, some of the technology cannot be classified as collaboration tools. Two examples would be registration and tracking features in a LMS and content software used to support specific courses. An example of the latter was the computer-based instructional and testing software used to develop students' baseline-knowledge in statistics before starting the statistics course in the Terry–PwC MBA programme.

10.4 Terry–PwC Consulting MBA programme

10.4.1 Brief Overview

Founded in 1912, the Terry College of Business is the flagship business school in the state of Georgia, one of the oldest business schools in the USA, and one of 13 schools and colleges at the oldest state-chartered university in the country. The Terry College is home to one of the nation's ten largest undergraduate programmes – with over 5500 students – and one of the most selective MBA programmes on any campus, public or private. *The Financial Times, Business Week, Forbes* and *US News and World Report* consistently rank its undergraduate and graduate programmes among the best.

With roots dating back to Samuel Lowell Price in London in 1849, PricewaterhouseCoopers (PwC) has a long history of providing consultation services. Currently, PwC Consulting serves businesses and organisations in more than 150 countries and territories with its over 150,000 staff people and 9000 partners. PwC Consulting's approach focuses on using multidisciplined teams that have a global industry scope and experience as well as knowledge of relevant issues and regulations. PwC Consulting's area of expertise spans 24 market sectors, grouped into three clusters: Consumer and Industrial Products; Financial Services; and Technology Info-Com and Entertainment. It is within this parent company environment that the PwC North American Consulting Group operates. We use PwC or PwC Consulting in this chapter to refer solely to the PwC North American Consulting Group with which the Terry College worked to establish the Terry-PwC MBA programme.

The PwC Consulting two-year MBA programme started with a proposal presented to PwC Consulting by the Terry College in March 1998. The goal was to create a programme that would provide benefits to PwC Consulting, its employees and Terry. PwC Consulting would gain a customised learning venue for highly valued employees that would encourage them to stay with the company, without removing them from the work force for two years. The employees would gain an opportunity to earn an MBA, obtain intense professional development without foregoing knowledge of current events in the company, and develop a network of relationships with other PwC consultants in the class. Terry would gain a broader knowledge and experience-base from an e-Learning programme linked with a large business consulting firm. The goal was to have approximately 40–50 students in each class.

A number of key features distinguish the Terry–PwC MBA programme and help make it successful. The programme is mission and outcome driven. The specialised curriculum supports and enables the blended delivery mode. The team and collaboration focus creates an environment where the sum of the parts exceeds the parts alone, which increases the amount of learning taking place and speeds it up. At the same time, the student workload management ensures that the students have time to interact on a regular basis to complete assignments and to successfully carry out their jobs. On-going feedback and evaluation mechanisms allow areas for improvement to be identified and effectively crafted. Finally, the support infrastructure and programme leadership creates an environment where technology is used for learning management and collaboration that enable successful learning to take place. These key distinguishing features are described in more depth in the following sections.

10.4.2 Mission and Outcome Driven

From the very beginning of the project, outcomes or goals of the two groups were made explicit and joint programme and technology mission

Mission: *To develop the highest quality MBA that meets the needs of PricewaterhouseCoopers and the MBA standards set by AACSB.*

Guiding Principles:

- Collaborative Relationship between Terry and PwC
- Customized and Flexible Program
- Best Teaching Faculty
- Relevancy
  - Leverage Participants' Experience
  - Apply and Integrate Learning into PricewaterhouseCoopers Environment
- Keys to Successful Learning
  - Collaboration (Dyad, Team, Class)
  - Discovery (Learning How to Learn)
  - Relevancy
- Effectively Balance Classroom and e-Learning Environments

Figure 10.3 Programme mission and guiding principles.

statements were developed. The programme mission statement and guiding principles are shown in Figure 10.3. The focus from day one was quality in terms of creating a programme that met PwC Consulting outcomes and Association to Advance Collegiate Schools of Business (AACSB) MBA accreditation standards. Unlike most MBA programmes that are exclusively designed by the academic side, the Terry–PwC MBA programme was developed through collaborative efforts of both parties. PwC Consulting personnel were involved in all aspects of the programme: course design, scheduling, selection and on-going management of students, etc. In designing the programme, our focus was on selecting the best teaching faculty, building relevancy in all courses, creating a collaborative learning community, and balancing classroom and e-Learning environments.

A good e-Learning technology platform was the key to achieving the programme mission and making the programme successful. Thus, we also developed a technology mission statement and guiding principles to keep our efforts focused. The mission and guidelines are shown in Figure 10.4. Our focus was on the learner and the learning process, not the technology. We wanted to create ways to use technology to support and enable learning. In 1998, e-Learning technology was still evolving. Thus, our focus was on tested and proven technology and for it to be effective it had to be compatible with PwC Consulting's technology infrastructure. Since both Terry and PwC Consulting were new to e-Learning, the project was viewed as a way for both parties to learn how to implement and effectively use e-Learning.

Each party had some very specific outcomes entering into this programme. PwC Consulting had three primary goals. Employee retention,

Mission: *To blend classroom learning environments with e-Learning technology to maximize learning experiences for PricewaterhouseCoopers employees.*

Guiding Principles:

- The focus is on the learner and learning, not technology.
- Learning is difficult, use technology as tools to support learning.
- e-Learning technology will be used both on and off campus.
- Use only tested and proven technology that is compatible with PwC technical environment.
- Apply and integrate learning materials and activities with PwC technology and work experience wherever applicable.
- Learn together how to implement and effectively use e-Learning

Figure 10.4 Technology mission and guiding principles.

their first goal, was key when the programme started but became less of a focus as time went on. Consultant turnover was a real problem in 1998 when the programme was being designed and initiated. PwC Consulting used this MBA programme as a way to retain key employees. Their second goal was to experience distance learning in a different way than they had implemented it in the past. PwC Consulting has extensive training systems. They bring students to locations to have face-to-face training and also have used distance learning. This blended MBA programme gave them some experience with a different design. Consultant development, their third goal, is still a key motivation in continuing this programme. Even with a current slowdown in the economy and with it layoffs in PwC Consulting, the company has kept a long-term view and therefore has chosen to continue this programme to expand the skills level and intellectual talent of their consultants.

The Terry College had a number of goals also. Two key goals were to become less dependent on state money and to explore emerging technologies as they relate to education. Terry wanted to make sure that they were ready for any changes in the way that education was going to be delivered in the future. This would mean a significant investment if they were to explore this on their own. The PwC Consulting programme provided funding support for startup costs and the development of an e-Learning infrastructure. After the startup of the programme, there would be revenue to invest into both this programme and other programmes. Producing new revenues meant that the Terry College would not have to react drastically to budget fluctuations at the state level. Additionally, the college wanted to gain a reputation in executive education and establish a valuable business connection.

10.4.3 Specialised Curriculum

Two years of study begin with a year of common courses covering core business areas tailored to PwC consultants' world and lead to specialisations in finance or marketing in the second year. PwC Consulting and Terry agreed to limit the second-year specialisations to these two tracks in order to facilitate PwC Consulting's outcomes. On the course level, content was also customised to the needs of the programme. For example, the typical core accounting content at Terry spans financial and managerial areas and takes place over two courses. This content was collapsed into one course tailored to a consultant. Though it was redesigned, the programme still maintains the academic core content to meet MBA accreditation standards. Terry's willingness to collaboratively customise their MBA was one of the key reasons PwC Consulting selected the Terry proposal.

10.4.4 Blended Delivery Mode

The programme vision was to blend face-to-face and distance learning for each course. This led to an initial agreement that face-to-face time for the programme's courses would equal 50 per cent of the in-class time allotted to a traditional MBA course. The remaining half of the traditional course time would be covered through distance learning while out-of-class study time usually expected from students would remain constant in either setting. A typical course like statistics would have 16 hours in the traditional classroom and 16 through distance media.

The first year of the programme starts with a two-week orientation and initial classroom sessions followed by three additional four-day weekend visits during the year. Thus, every three months the professors and students touch base and participate in any activities requiring the information-rich communication channels available during face-to-face learning time. The second year starts out with a one-week session followed by four four-day weekend sessions evenly distributed during the year.

For the faculty who were to teach in the programme, all new to distance learning, the 50–50 rule presented a problem: how to transfer a course that has been taught for years face-to-face into a mixed or blended context was a major issue. There were different approaches to this problem. One of the first approaches was to teach in the same order, as always, and when it was time for the distance portion, try to fit the course at that point into the technology. This approach worked well for courses that built on previously taught material, like accounting.

Some faculty would teach all major concepts in on-campus time, and during the distance portion, they would have the students practice what had been demonstrated. This approach worked well for statistics. The students had the big picture idea of what was going on, and it was later

cemented when they practised it. With this approach, the face-to-face time is leveraged.

Some faculty would completely rearrange their course so that activities could be done during the face-to-face time with little lecture, leaving lecture for presentations while at a distance. This worked for operations and legal courses. Another approach was to leverage some of the distance time. The ethics class started out with lecture and tried to get the students ready for some ethical conflicts that would be handled at a distance. These ethical conflicts required thoughtful time before advancing and were therefore perfect for the distance portion. The key to effectively blending the courses ended up being that each mode, either distance or face-to-face, had advantages and disadvantages, and the course would have to be designed to take these into account. As we learned how to be successful in each course, we adjusted the 50–50 rule giving some courses more face-to-face time and some less. However, we are not experts in this area. There is a great need for research and sharing of practice on how to effectively blend different approaches and technologies.

A variety of technologies are blended to support both on-campus and distance portions of courses. Additionally, student teams learn a variety of collaboration technology options to meet their needs. These technologies range from e-mail and instant messaging services to team information spaces. These technologies have been the key elements in individual teams' success. All technologies are used when students are on campus and when they are working from a distance. These technologies will be discussed in more depth in Section 10.6.

10.4.5 Team and Collaboration Focus

The programme designers aimed to retain the importance of collaboration, especially through teamwork, that characterises successful MBA programmes. Research has shown that collaborative learning is very effective, when done well, in college environments (Johnson et al., 1998). In addition, research has shown that the most successful distance learning courses were with class sizes of 15 or less. A smaller class is much easier to facilitate through technology especially when maximum interaction is wanted. Since the Terry–PwC MBA programme was looking at a class size of around 50, the effective use of learning teams and collaboration technology was imperative.

Thus, the Terry–PwC programme focused on building strong virtual learning teams of 4–5 people, where much of the initial discussion and work is done within each team, then between teams or between teams and the instructor. A great deal of time and effort is expended to ensure the development of effective virtual teams and the development of the overall class as a learning community.

Most of the first week in the two-week initial on-campus visit is devoted to team/class development. This is done through a team skills course that focuses on the effective development and facilitation of virtual teams. During this class, teams establish mutual goals and mission statements, work out a code of conduct that each member is willing to support, and develop a media plan that outlines how the team is going to use collaboration technologies to work together. It was our feeling that for a virtual learning team to be successful, the team needed to have a good technology platform that members are committed to! The development of the media plan facilitates commitment.

10.4.6 Student Workload Management

A balance had to be struck between the demands of maintaining the students' jobs and attending college. A typical workweek for consultants is 60–70 hours, and the MBA programme might demand up to that many hours during peak times. As a result the Terry faculty agreed to limit study time during off-campus periods to 15 hours per week, and PwC Consulting limited work time to 40 hours. This joint rule on timesharing was monitored through feedback mechanisms. Although there were continuous problems on both sides implementing this rule, this joint timesharing rule was critical to the success of the programme.

10.4.7 On-going Feedback and Evaluation

Regular feedback is built into the programme. Every week students send in anonymous online feedback focusing on their experience in the programme. While this feedback includes more discrete items such as time spent on assignments and on PwC Consulting project work, it also gives them a chance to identify any problems they are facing. The programme faculty and administrators receive this data and review it weekly. As issues such as assignments far exceeding the specified number of hours arise, solutions are crafted such as using an on-campus MBA student to benchmark the assignments or redesigning the assignments. In one case a professor, after viewing his own assignments benchmarked, decided his course should occupy a three-hour time slot instead of two. Course and programme changes are often based on this feedback.

During on-campus visits especially early in the programme, feedback was gathered daily to ensure we were doing the right things. These surveys were done using LearningSpace. This survey and assessment functionality is critical to e-Learning and common in most LMS/LCS. We would never have been able to continuously improve the programme without this type of collaboration technology to gather input from 40–50 students.

10.4.8 Support Infrastructure

Our research into e-Learning had clearly shown that without adequate support, the programme was doomed to failure. Thus, providing support was a guiding principle from day one. When the programme was launched in October 1998, the programme had three full-time employees and nine part-time MBA graduate assistants supporting the faculty and students. They were referred to as the Distributed Learning Group (DLG). The three full-time roles consisted of a Lotus Notes Administrator, a Faculty Liaison who focused on effective use of technology by faculty, and a Technical Support Assistant, a person who provided backup to the Notes Administrator and did general technical support. This support team provided initial technology training and then served as an on-going technology help desk providing support throughout the programme.

One of the key issues that arose in developing support was bridging the competence gap. First, we had to get technical staff trained in Notes, LearningSpace and other technology tools, then train faculty and students. While some of the faculty and students tended to develop into tinkerers/experts interested in playing with and innovating their own uses of technology tools, others tended to be "good soldiers"/novices who expected to be told the right things to do, how and when.

The support infrastructure had to be designed with enough flexibility to accommodate both of these user types. The PwC Consulting programme achieved this by providing initial support through training and weekly meetings that mixed the support staff with the users and focused on specific technology topics. These meetings addressed such topics as the scheduling tool, instant messaging, or developing presentations with audio, depending on the faculty's requests.

Programme leadership was also a critical issue in support infrastructure development. In a joint programme like this, we needed strong leadership from both Terry and PwC Consulting. Our leadership team consisted of three main players. From the UGA side, one faculty member who had extensive technological knowledge and tenure in the university system served as programme and project manager for technical and curriculum/content issues. The knowledge and experience of this person served to garner the respect and open channels between technical staff, faculty and administrators. The Terry MBA Director managed the administrative aspects of the programme.

PwC Consulting had a person who coordinated all activities from the PwC Consulting perspective. Initially, he was devoted full-time to this programme and once the programme was implemented it was still a major time commitment. Once the contract was signed, PwC Consulting also designated a partner, who served on the Executive Committee, to oversee the project. The coordinator could use his access to this top management person, if needed, to facilitate PwC Consulting's support for the programme.

Utilising his staff and other PwC Consulting resources as needed, the PwC Consulting coordinator is involved in all aspects of the programme: administration of contract and all travel arrangements for students, programme/course design, selection of students, and on-going management of student issues such as violations in project time commitments. Students are first nominated by partners and selected by PwC Consulting management. Then students have to be accepted by the Terry College Admissions Office to get into the MBA programme. Without this strong leadership commitment from PwC Consulting, the programme would not have been a success.

The strong leadership commitment from the Faculty project leader and PwC Consulting project coordinator, their ability to effectively work together and to mobilise resources in both of their institutions were critical to the development and success of this programme in the early phases. Any university or business looking at developing such a relationship needs to clearly focus on building an integrative leadership structure to develop and manage the programme.

10.5 Implementation Process

The first year of the programme ended in October of 1999. Getting there required addressing seven critical stages: pre-planning; planning; technologies implementation; faculty training; starting the face-to-face component; starting the distance component; and on-going evaluation. The complete project start-up time line is shown in Figure 10.5.

10.5.1 Pre-planning

While our discussion is based on the time period from when the contract was signed, some key faculty at the Terry College had been investigating e-Learning technologies prior to this period. They had done research into how different technologies could be used in an e-Learning environment. While they did not have a definitive system in mind, they were able to pick a couple of products that seemed to show the most promise. Thus, at the point that PwC Consulting and Terry got into serious discussions, there was not only expertise in-house, but there was also a conceptual framework already forming in the mind of the faculty member that ended up being a key lead for the project.

The contract itself, signed in March 1998, ended up answering many questions as to how the programme was to be delivered giving the planning phase some boundaries to work within. Time would be close to 50

<table>
<tr>
<td>

October 1997

1. *Pre-planning*
   - Develop Program and Technology Missions/Visions and guiding principles
   - Alternate technologies evaluated
   - Feasibility assessed
   - Program proposed February '98

March 1998

- Contract given to Terry

2. *Planning*
   - PwC MBA curriculum approved
   - Support staff hiring
   - Training planning

June 1998

3. *Technologies implementation*
   - Infrastructure design and installation
   - Faculty use drives modifications
4. *Faculty training begins*
   - Weekly face-to-face meetings
   - Feedback loops help ensure that training is effective

</td>
<td>

October 1998

5. *Students arrive for two weeks on campus*
   - Orientation, training in technology and team skills
   - Teaching of core courses begins
   - Feedback gathered daily, adjustments made
6. *Students begin the distance element*
   - Students work fulltime and travel
   - Feedback leads to personalized support and faculty discovery of best practices
   - Faculty and support staff meet regularly to make adjustments and share best practices

January 1999

7. *Students arrive back on Campus* for first weekend. Steps 6 and 7 are repeated two additional times in first year

October 1999

8. *Students return to campus for seven-day stay* to kick off their second year. Steps 6 and 7 are repeated for 4 weekend stays during second year

October 2000

First Class graduates!

</td>
</tr>
</table>

Figure 10.5 Project start-up timeline.

per cent on campus (compared to the number of classroom hours for a traditional MBA) and 50 per cent off. There would be 40–50 students per class. Each class would take two years to graduate and a new class would start yearly. Classwork would be scheduled 50 weeks a year with two weeks off at Christmas. The first class would start in seven months, October 1998.

The state of the technologies gave some other boundaries. Dial-up was the standard way for students and faculty to get to the Internet from their homes or hotels. Learning Content Management Systems were adequate but did not have a real programme feel and there would have to be some customised solutions implemented.

Pre-planning played a critical role in the programme's success. Once the programme's vision, outcomes and guiding principles were established, feasibility had to be ensured. Feasibility analysis has economic, technical and organisational components. From an economic standpoint, analysing the cost of implementation and on-going support led to an understanding of the budgets that would be necessary. On the technical side, various software programmes and their hardware requirements were researched. This technology was then compared to the existing infrastructure to identify additional investments needed and to understand how the learning content would have to be designed and blended to use the technology effectively. For example, since Lotus Notes already existed as a platform in PwC Consulting, looking for a Notes-based system that met the programme's requirements became a high priority.

One key to the organisational analysis was creating a blended delivery design for the programme. Faculty workloads and attitudes about participation and commitment levels were examined. Many faculty members and PwC Consulting staff involved expressed deep reservations about the learning quality in an entirely virtual programme. Yet this was essential to the programme, as the PwC Consulting participants would have to continue working off-campus while studying in the programme. The blend of face-to-face and virtual delivery served both organisational needs.

10.5.2 Planning

Once the contract was signed and the deadlines were real, planning started. Who was going to manage the programme all the way down to who would make the coffee when the students were on campus had to be decided, and people needed to be appointed or hired. At this time, the programme started hiring the new infrastructure support group, the Distributed Learning Group (DLG), which would then purchase and implement the specific hardware and software packages necessary for the programme. The first two staff members were hired and working as of June 1998. At the same time, key project leaders handled the logistics for conducting faculty training and student orientation. Specific tasks ranged from setting up a faculty listserv and discussion databases to allow faculty to share experiences, problems, and solutions, to developing training materials that would cover primary functional areas for the faculty. The Terry project leader working with the MBA faculty committee developed the curriculum and got it approved by the MBA committee and the Terry faculty.

10.5.3 Technologies Implementation

Once some of the basic structure was put into place, such as servers and server and client software, the DLG explored how they could be used to fit the needs of the students and faculty. As deficiencies were found, new technologies were researched or developed and then implemented. This ended up being analogous to building a house with a general plan and working out the details as it gets built. While this does not seem to be the most logical way of handling the implementation of a professional programme, both the Terry College and PwC Consulting entered into this agreement to learn more about e-Learning. This also set the mindset for the programme for the first year: plan, develop, implement, evaluate, and adjust. This was done at every level from the director down to the student.

Since the technologies were to be managed internally, the new DLG staff had to obtain training on them and install them. They spent many hours debugging and configuring the necessary servers and networks, setting up workstations for faculty, and establishing procedures for managing the on-going needs of the programme. As one of the original staff members put it, "Don't underestimate the need for establishing procedures." These procedures then allowed the hiring and training of new staff with less of a time lag to explain tacit procedures. This was particularly important because DLG used a lot of part-time students.

Technologies implementation involved checking for not only physical linkages and functionality of the system but also refinements to the design of the architecture. As faculty feedback began to arrive during weekly and sometimes more frequent meetings in July and August, the support staff worked continuously to derive the maximum value from the LearningSpace software and other software chosen for the distance elements of the programme. For example, a programme-wide metacourse was designed as an improvised tool for programme-level interaction and management.

10.5.4 Selection and Training of Faculty

From our research on distance learning, we knew that the faculty/trainer role was key. We decided to go after the best teaching faculty we had in each area regardless of their technology background. We felt that good teachers would have the best skills to use technology effectively both in the classroom and from a distance. We knew we could teach them to use the technology, but it would be much more difficult and time consuming to teach them how to be good teachers.

Technology readiness varied greatly, from a few faculty who had just started using e-mail and were somewhat familiar with word processing to very technology savvy users. The training dealt with this by covering a variety of technology options, appealing to the knowledgeable users, while also including a four-week tutorial on core functions of the LearningSpace software and how to apply them to each individual's content, appealing to the novices.

Faculty training began with a focus on the functional areas of their teaching and how specific technologies could fit specific tasks in each area. These functional areas came from their standard classroom teaching and included: how to lecture online; how to run a discussion online; how to assign and collect papers online; how to administer exams online; how to hold office hours online; and how to facilitate student teamwork online. Some larger issues also arose, such as how to integrate the distance component of a course with the face-to-face time. This issue was a source of anxiety for many, but ultimately required simply getting into the semester and experiencing the solution appropriate for each course.

The feedback and collaboration between faculty during and between sessions, through the listserv, discussion databases and during weekly meetings also led to dynamic adjustment of the training when issues arose. Toward the end, this training process had created a working team composed of faculty and support personnel who could then help each other respond to challenges during the first semester.

10.5.5 On Campus Learning Begins

On 17 October 1998 the new students arrived on campus, and the face-to-face component began. These two weeks were intense. Students had to master initial content, the technology, and become oriented to each other in a new environment. Several sessions focused on developing and facilitating learning teams and student relationships. Research has shown that these relationships are valuable in maintaining a personal sense or feel during distant interaction. The personal feel relates to the individual's or team's sense of information richness within the distance technologies. Face-to-face conversations contain the richest information, full of feedback capabilities, multiple cues, and a personal focus. Research indicates that students who have a prior face-to-face sense of each other and trust intact can move more easily into and use online learning environments even if they are using tools that do not offer as rich an information channel as face-to-face interaction.

As such, prior face-to-face interactions affect the effective use of technologies. Groups having these experiences report more comfort using less rich information technology, such as audio and e-mail instead of video and synchronous chat. Forging personal connections requires information richness in a variety of personal details. Thus, this activity seems best handled in face-to-face environments. Trying to make personal connections happen effectively in a totally virtual context can lead to a larger investment of time and energy. Thus, the first two face-to-face weeks served the typical course role of beginning content delivery along with a more critical role of developing the relationships and teams which would enable the transition to distance learning.

10.5.6 Distance Learning Begins

When the two-week campus visit ended, faculty and students had to rely on the distance technologies for the first time. Professors would record lecture audio tracks synchronised with on-screen slide presentations. These bundled lecture files would then be uploaded to their course in LearningSpace. When students logged on their computers they would automatically download all of the new content files and upload all of their work, through the replication function in Lotus Notes. Replication allowed them to continue studying and interacting with the class even when disconnected from the Internet. For example, students could be on a plane between consulting jobs, listening to lectures and reading and responding to the latest updates in the online threaded discussions. Arriving at their destination they could dial into the Internet again and touch base with everyone else's updates. They could study disconnected without being disassociated and depersonalised.

10.5.7 Evaluation and Feedback Lead to Improvements

Throughout the semester students sent in feedback to their professors and to the programme staff. When classes ended, the students filled out evaluation forms. The suggestions from feedback and evaluations helped to craft programme improvements. Students could compare and contrast the formats of the different courses they were taking and helped to spread knowledge of best practices from professor to professor. The professors would then meet and share tips and techniques. On the support staff side, procedures for keeping the technology running smoothly improved. For example, we learned that due to the dial-up connections that some students used, lecture files needed to be limited to chunks of about 20 minutes or less, or about 1.5–2 Mb. We also found that short lectures from a distance were a lot more effective from a learning perspective as well.

10.6 Collaboration Technology Used in the Programme

The Terry College looked at a wide range of technologies in an effort to find a good mix for a blended programme. We looked at the benefits and drawbacks, trying to leverage strengths and minimise weaknesses. In making our technology selections for the programme, we used the two models introduced earlier in this paper – the time-space technology matrix (see Figure 10.1), and the blended learning environments (see Figure 10.2). The programme focused on blending the three environments outlined in Figure 10.2: online anytime, online live and classroom.

This section overviews the technology used in each of these three environments.

10.6.1 Online Anytime

Online anytime technology allows access to information anytime, anyplace. It is usually the core technology for most e-Learning efforts. The primary technologies used in this area are shared information spaces and e-mail. For the Terry College this meant that lectures, handouts, reading material and classroom conversations would have to be digitised, and then these electronic documents could be visited by the instructor and students when and where it was convenient. The initial faculty and support investment in teaching a course would be high, but following offerings of the same course would already have most of the content ready. Research has indicated that initial course development work takes about three times more for an online anytime course. Having the courses in electronic format also meant that the course could be easily updated on the fly.

A major drawback for Terry was that faculty did not know how to develop this type of instruction for this learning environment. This is why faculty training was such a key issue for the programme's success. For PwC Consulting the advantages were that students could work on the courses during off worktime hours. They did not have to be transported from their work, and they did not have to miss work. Since PwC Consulting employees already used this type of communication technology (Lotus Notes), no additional PwC Consulting infrastructure was needed. This also facilitated student training.

Since the programme was going to be a blended distance and face-to-face education, there had to be a mix of technologies. It was decided early on that there would need to be an overarching technology that would hold both distance and face-to-face course materials. This would allow the students and faculty to access and post information to a common space. We briefly introduced this type of technology earlier as a Learning Management System (LMS) or Learning Content Management System (LCS). We chose a LCS from Lotus called LearningSpace. In looking at the time and place model, LMS/LCS fits primarily in area IV, different time–place, although they can be used to support all three of the other areas in the time-place model.

We started with LearningSpace version 2.5 and have moved to version 3.0. Since the programme began, Lotus and IBM have developed newer versions 4.0 and 5.0. These latter versions took more of a commercial or corporate focus and web-centric approach. Many of the key academic features, such as electronic assignments, were dropped in version 4.0. In addition, the Lotus Notes infrastructure was replaced with a server-side

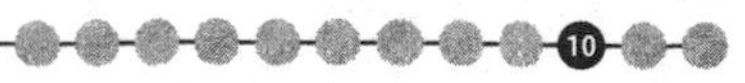

database one. This meant by upgrading to version 4.0 we would lose the replication feature of Lotus Notes, forcing students to be connected to the Web to use the system. Based on these concerns and others, we decided to stay with version 3.0.

Since PwC Consulting used Lotus Notes for all the consultants in the programme, LearningSpace, which is a Lotus Notes application, was a natural fit. One advantage of LearningSpace over some of the other LCSs was its ability, as noted earlier, to replicate a complete copy of all learning materials on student computers. Replication allows students to work on course activities while disconnected from the Internet. For the Terry College, this had the additional benefit of decreasing the reliance on their network. Concern about network outages would not be as critical because students would have a copy of the course on their laptops.

Each LearningSpace course has five Notes databases, four that the student has access to and one that is for the faculty member to manage grades and make tests. The first four are the ones the student uses directly:

- *Schedule*: contains the syllabus for the class outlining learning activities with links to all the associated support materials.
- *MediaCenter*: this is the electronic library for each course that holds all handouts, articles, recorded lectures, etc. with links to other resources.
- *CourseRoom*: roughly equivalent to the classroom. It provides for threaded discussions and is where assignments can be handed in by the students and handed back by the faculty member.
- *Profiles*: this holds student information, pictures of the students and student portfolios of grades.
- *Assessments*: this database holds quizzes, tests, and surveys along with the grade book for the faculty member. Students complete assessments by accessing them from the Schedule database.

Figure 10.6 shows the Instructor LearningSpace interface to the Schedule database for the Teams Course, which is the first course taken in the programme. The student interface would be the same except they would not have a tool bar for creating and editing entries. The interface is a typical Lotus Notes interface with a view menu on the left-hand side and documents listed in the centre of the screen. LearningSpace is available through a Lotus Notes client or a web browser. The Notes version, pictured in Figure 10.6, has more functionality in it and must be used to create documents, but both interfaces look similar.

The Schedule database captures the course syllabus as a collection of organised Notes documents that describe learning activities. The organisation for this class was by module but could have been by week or some other organisation type. For example, in module 6, there are four Notes

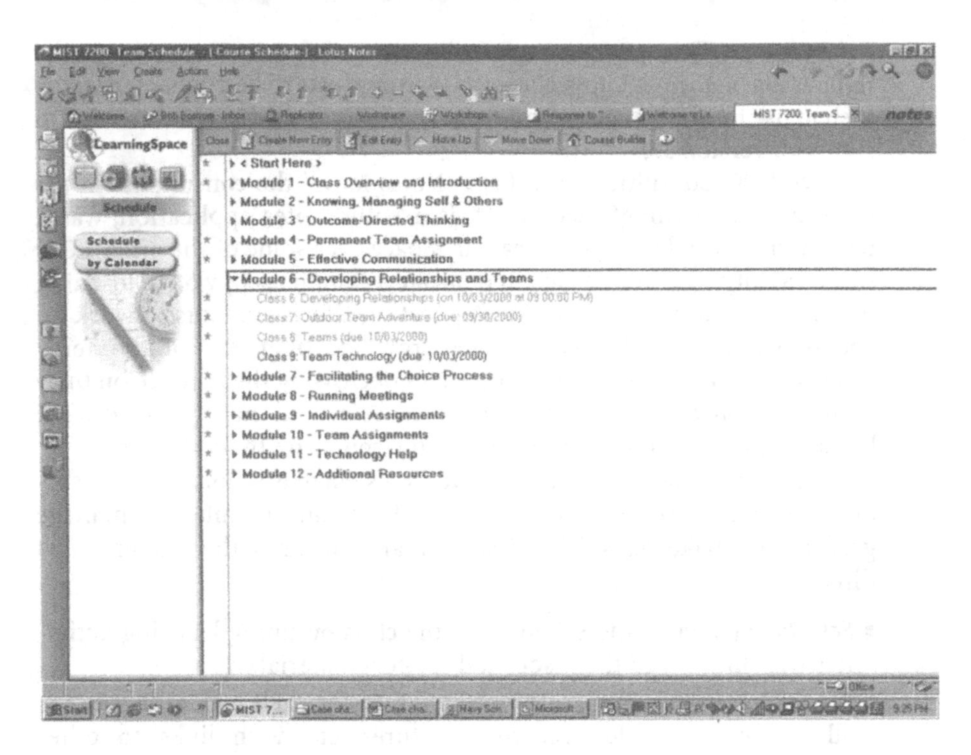

Figure 10.6 Schedule database for team skills course.

documents describing four different classes. The student would simply click on a document, "Class 6:...", to get information for that class period. The student can see learning activities displayed in calendar format by clicking on the "by Calendar" button. The student can easily switch to different databases by clicking on icons just below "LearningSpace" in the left-hand corner or through links embedded in Schedule documents. (For example, the "table with people around it" icon designates the CourseRoom database. Clicking on it will move you to the CourseRoom interface.)

The following scenario outlines how students would use LearningSpace and other tools to accomplish a learning activity from a distance:

- Read assigned readings as outlined in *Schedule database entry* (readings in textbook or accessed from *MediaCenter*). Students listen to mini lecture with slides when desired, accessing session from *MediaCenter*.
- Students *post a note in a discussion in CourseRoom database.* Instructor or classmates respond to questions or ask for clarifications. Instructor or students may create a most *frequently asked questions discussion in CourseRoom* for students.

- Team or class discusses readings using a *discussion in CourseRoom*. Certain students or a team may be asked to start discussion by reviewing/critiquing a reading. Students or teams may complete a homework assignment related to the readings (this may be one the student creates from their own work environment). Teams will use their *TeamRoom* shared information space to work on homework from a distance. They may supplement this with real time communication (*ICQ, NetMeeting*) and phone conferences.

- Assignments will be electronically submitted (*submit as a LearningSpace assignment in CourseRoom*) to professor for grading. The instructor uses special instructor facilities in *CourseRoom* to grade assignments. Graded homework will be automatically returned to students (assignments posted in *CourseRoom* as graded), and their grade will be posted to the *electronic gradebook in Profiles and Assessment databases*. Students may also work through some self-assessment questions or exercises to check their understanding of material. These may be purchased or developed instructional programmes that would be launched from *Schedule or a LearningSpace assessment* that was developed by the instructor using facilities in the *Assessment database*. Students may also be required to write an entry in their electronic journal that integrates and applies knowledge gained from learning activities to their work situation. The journal could be implemented using a *CourseRoom Discussion* that keeps specific entries of each student separately, allowing access only to personal entries and overall access to the instructor *or by using a separate Notes database*.

Finding an effective workspace for learning teams was a problem for us. LMS/LCS usually offers some team support workspaces and facilities but these are usually limited and, thus, additional technology is usually needed. Although LearningSpace handles team assignments and grading, it does not provide any team support technology.

However, Lotus offers another product, TeamRoom that met our needs. Since TeamRoom, like LearningSpace, was a Lotus Notes application, it was easy to move documents between TeamRoom and LearningSpace. The class was divided up into teams, and these teams were used in all classes. This team approach was a central focus to the programme and needed a technology to support it. TeamRoom was slightly modified for this purpose, and it gave a place for teams to meet in a virtual space (using discussion databases) and a place for them to share documents and references. We modified TeamRoom to hold additional team development material such as team purpose and outcomes, code of conduct and media plan. TeamRoom also allowed the creation of information spaces for subteams and the scheduling (due dates, etc.) and assigning of responsibility for course assignments.

10.6.2 Online Live

Online live technology primarily supports real-time communication while communicators are in different places. We trained faculty and students in the use of audio-video conferencing (NetMeeting) and instant messaging (ICQ). The advantage for the Terry College was that faculty could still lecture to or hold office hours with students but not be required to be at a certain location. This meant that faculty could teach while at a vacation home or while overseas at a conference. The College could offer the best faculty for the programme while offering the faculty member the flexibility to pursue other interests. For PwC Consulting, students could be on any project and would not have to spend time travelling to one location. Students could also use this technology to support their teamwork from a distance.

The first drawback was the coordination of a common time for faculty and students to meet. During the time of any one course, faculty and students were in many different time zones, some even overseas. Finding a time that was not too late and when all concerned were not at work was difficult. Some faculty did work out electronic office hours that used these tools with regular phone contact if needed. Students used instant messaging as a way to stay in contact with each other and for quick questions and file transfers to team-mates.

Most of the learning teams used audio conference calls through a phone bridge once a week to coordinate team activities. PwC Consulting, not Terry, supported these conference calls. Teams were organised so that team members were in the same or adjacent time zones wherever possible to facilitate real time meetings. The remainder of the time they coordinated work through the shared information space technology, TeamRoom, described above. All team-mates were usually connected into TeamRoom during their conference call to access and share relevant information and files.

The other more critical drawback at the start of the programme was the state of the Internet. High-speed connections were not readily available in residential settings or in hotels. This has been changing and we are thus actively encouraging the use of more online live tools and investigating the addition of new tools, especially a virtual classroom tool such as Centra, WebEx, Lotus Sametime or Horizon Live. Virtual classroom tools have added playback facilities that allow a student to review a session or experience it if they could not attend. The playback feature deals with the problem of getting people together all at the same time. These tools are being used to effectively train students in both academic and corporate environments. We feel they will also be excellent tools to support our learning teams. By the fall of 2002, we hope to use a virtual classroom in the Terry–PwC and other MBA programmes.

10.6.3 Classroom

Classroom learning happens in the same place and the same time. We explored technology that would support a classroom setting or a face-to-face meeting. We also wanted students to have available and use the same technology when they were on campus that they used from a distance.

Our challenge in this area was to get faculty to use technology as a "digital surround" to support classroom activities. Since all students had laptops and brought them to class, shared information spaces and tools could be accessed and used. For example, LearningSpace surveys or instant messaging chats could be used within the classroom to support case discussions and engage students in other ways. Some faculty used collaboration technology in this way, but it was limited. Most of the time technology was only used to access class resources in LearningSpace. It was limited for two reasons: we did not focus on the technology's broader uses in our training, and we needed better tools to support this, such as electronic meeting support tools or virtual classroom tools. We are currently exploring integrating these types of technology into our infrastructure.

Area II in the time–place matrix, different time and same place, is usually not addressed in e-Learning discussions. It implies a common physical place where students or faculty could leave things for each other. We have this capability when students are on campus since they are always in the same classroom and materials are left for students there. However, our focus was not on a shared physical place, but rather a focus on shared information or virtual spaces that could be shared at the same or different times. Thus, even if one is in the same place, such as the classroom, we use collaboration technology to enable people to share information through virtual spaces.

10.6.4 Different Levels Involved in e-Learning

One of the biggest problems encountered using LearningSpace for the programme was its sole focus on the "class" versus the larger MBA programme. The original design of LearningSpace did not allow for a programme view. While each class is important for the student, it is the programme or how each class fits with the others that make an MBA degree valuable. As a solution, another LearningSpace course was created to provide this programme view, but this solution still lacked many important features such as an easy way for the students to see an overall picture of what was due for them in any one of their classes.

This problem highlights the point that there are multiple levels that you need to support with e-Learning technology. Most organisations

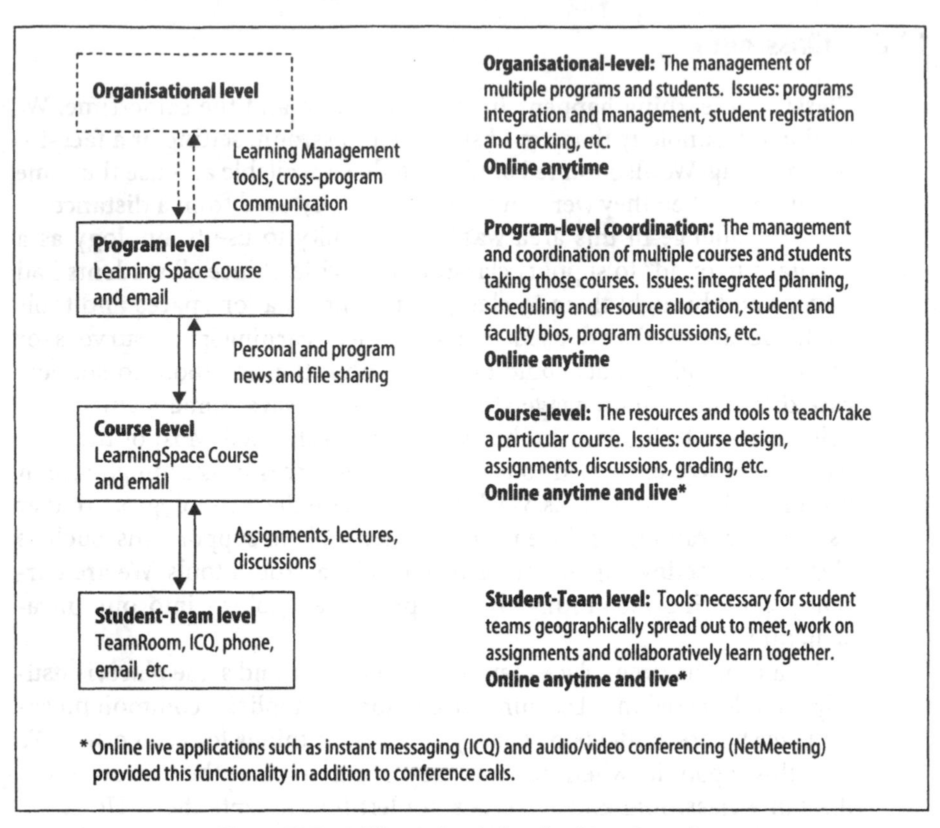

Organisational level
Learning Management tools, cross-program communication
Program level
Learning Space Course and email
Personal and program news and file sharing
Course level
LearningSpace Course and email
Assignments, lectures, discussions
Student-Team level
TeamRoom, ICQ, phone, email, etc.
Organisational-level: The management of multiple programs and students. Issues: programs integration and management, student registration and tracking, etc.
Online anytime
Program-level coordination: The management and coordination of multiple courses and students taking those courses. Issues: integrated planning, scheduling and resource allocation, student and faculty bios, program discussions, etc.
Online anytime
Course-level: The resources and tools to teach/take a particular course. Issues: course design, assignments, discussions, grading, etc.
Online anytime and live*
Student-Team level: Tools necessary for student teams geographically spread out to meet, work on assignments and collaboratively learn together.
Online anytime and live*
* Online live applications such as instant messaging (ICQ) and audio/video conferencing (NetMeeting) provided this functionality in addition to conference calls.

Figure 10.7 Different levels involved in e-Learning.

would have the four levels described in Figure 10.7. Using the Terry–PwC MBA programme as an example, we have already discussed the individual team and course levels. Students work individually and in teams within a given course. Technology was needed to support both individual team and course levels. We also needed support for overall management of the programme in terms of integrated scheduling and programme-wide discussions, for example. Similarly, from an organisational perspective, it would be useful to have technology support the management and integration of different learning programmes. For example, the PwC Consulting MBA is one of five MBA programmes in the Terry College. Terry continually has problems with scheduling, resource allocation, sharing information, etc. between these programmes that e-Learning technology could support. In an organisational setting, the organisational view would most likely be managed by the Human Resource area as it schedules and tracks employees through the company's many training programmes.

The differences between LMS and LCS, discussed earlier, are highlighted using the level model in Figure 10.7. A LCS usually focuses on course and student-team levels, while a LMS tends to provide support for all levels. Most LMS/LCS do not provide online live support or good learning team support tools. However, the trend is to provide these capabilities by allowing seamless links from LMS/LCS to other tools. In our situation, we use LearningSpace as our core LCS. We created some limited ways to support programme and organisational levels as discussed above, but we could have used much better support at these levels. In addition, as we outlined above, we had to add a number of technology tools to support live collaboration and teamwork. LearningSpace 3.0 did have an online live module but it did not perform well enough to meet our needs. It is our understanding that the newer versions of LearningSpace have improved the online live capabilities.

We are continually monitoring the evolution of the LMS/LCS market to plan our e-Learning strategic direction. It is clear that the trend is for more and more seamless integration of tools using a LMS/LCS as the core online anytime technology. This integration is being greatly facilitated by the development of standards allowing reusability of content between courses from different vendors and interoperability of technologies.

Although not directly related to collaboration technologies, this issue of learning content is very important to a company's e-Learning strategy. Let's look at it from the level's perspective, Figure 10.7, and use the PwC Consulting MBA programme as an example. You can find digitised learning content everywhere. When we started this programme, not a lot of content was available so faculty developed a lot of content on their own and incorporated pieces developed by others. We are still doing this today. However, today there is a lot more electronic content available to plug into a course! The trend is toward the development of learning objects and complete courses that use these objects. The objects are reusable and easily moved between different technologies and vendors because of standards. For example, we are now looking at buying learning objects and complete courses to use, much like we have done with textbooks for ages. There are now companies that sell core MBA courses. Clearly, at the organisational and programme levels we need not only consider the development and management of e-Learning and digital collaboration tools and infrastructures, we need to develop strategies for acquiring, developing and managing learning content. In 2002, many people are struggling with this content issue including most universities.

10.6.5 Programme Success

By the end of the first year, though there had been plenty of bumps and challenges, students reported very high satisfaction and professors could objectively see that successful learning had taken place. Some representative student comments from the first year of the programme, captured

- I like how this program respects us as established business-people and seeks to have us find the relevance in what we do to the course materials. I like the parallels that I have found amongst all the courses, which is something I certainly had not experienced to this degree when I was in undergraduate school. Obviously, the technical support is outstanding, and the level of enthusiasm and interest-holding that the professors have is top-notch.
- The program is a great opportunity to broaden my knowledge base and a chance to further my career. The program needs to keep responding to our concerns and needs. If that occurs then we together can make this a successful program.
- The teamwork aspect was definitely something that is not only going to make the program special, but also bearable. It was good to be able to "bond" with the people that are going to go through the same trials and tribulations over the next two years.
- I developed some great relationships. I will not only take those relationships 'home', but I hope I carry them through my life. The on-campus visit let me develop a great excitement for this program. I am excited to work with the technology and be at the front of distance learning. I hope to keep this excitement for the next two years.
- I am completely amazed at how well everything has been going. We are being given the best professors that I have ever encountered and the technical support is incredible. I hope the workload slows down just a little, otherwise, keep everything as it is.

Figure 10.8 Representative student comments from the first year.

from weekly student feedback, are shown in Figure 10.8. Blended delivery, enabled by e-Learning technology, made this possible.

The success of the overall programme lies in its fulfilment of the original goals. Terry aimed to deliver quality learning, using a blended approach, and to build an effective e-Learning infrastructure. Based on input from all stakeholders, Terry has accomplished these goals. Two of PwC Consulting's goals were employee retention and development. PwC Consulting feels these have been accomplished. Students continue to give high praise to the programme. In 2000, PwC Consulting renewed the contract for another six years. During the first year, we also had two professors from the UGA College of Education evaluate the programme, giving it high marks (Schrum and Benson, 2000a, 2000b). The latest accolade for the programme was it being rated one of the top online graduate programmes by *US News and World Report* (Special Report: E-Learning, 2001).

Another measure of success is retention. Distance learning research has documented high dropout rates (numbers have ranged as high as 30–80 per cent) and high dissatisfaction with distance delivered classes. We discussed the high student satisfaction with the programme previously. In terms of retention, two classes have graduated to date. Of the 91 who started the programme, 89 graduated. Two dropped out because they did not like the format, especially the distance component. Two others had to drop out toward the end of the first year for personal reasons, but they both resumed the programme with the second class and graduated. Three of the graduates have gone on to become partners in PwC Consulting.

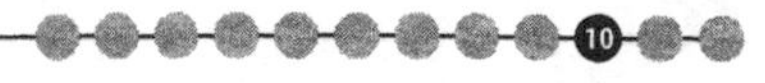

10.7 Key Implementation Guidelines

How do we explain the success of the Terry–PwC Consulting MBA programme? This last section will summarise lessons learned from the case discussed as a series of guidelines for those wishing to implement e-Learning programmes. The guidelines or principles are supported by brief summaries of the case material. Although these guidelines are derived from an academic setting, they are applicable to both organisational and academic environments. Many of the change guidelines are supported in the academic and practitioner literatures. For example, see Kotter (1995) and Johnson and Blanchard (1998) for good practitioner references, and Cummings and Worley (2001) and Bostrom (1983) for good academic references.

- *There must be a felt need or urgency for change*

  The research literature concludes that for successful organisational change to take place a felt need or urgency for change must be clearly established. This means that the need must exist and key stakeholders must be aware of it. The felt need may be existing problems that need to be solved or opportunities that are worth pursuing. In this case, both Terry and PwC Consulting had a number of goals, reflecting critical opportunities, which created a strongly felt need for both parties.

- *Involve the right people and make sure you have a coalition/leadership powerful enough to make change effort happen*

  The change literature talks about involving the right people in the change effort and that some of these people need to be powerful enough to make the change happen. The latter group is often referred to as change champions or a guiding coalition. The Terry project leader and the PwC Consulting project coordinator were the first guiding coalition. The Terry project leader was a MIS faculty member who was going to teach in the programme, chair of the MBA programme committee at the start of the programme, someone who had foundational knowledge in e-Learning and collaboration technologies, and someone who did a lot of consulting as well. He certainly had the right attributes to lead this project.

  The PwC Consulting coordinator was an ex-academic with a PhD and, thus, understood the academic world well. He was also in a powerful enough position in PwC Consulting to make things happen. The skills of these two project leaders and their ability to work together and to mobilise resources in both of their institutions were critical to the development and success of the programme.

  Once the core group of faculty was selected, the guiding coalition changed to include this faculty group. Clearly, selecting the best teaching faculty and getting them to work as a team was critical to successful implementation. The project leadership and the programme faculty continue to be the guiding coalition behind the Terry–PwC MBA programme. In addition, the hiring of excellent support staff, especially technical support, was and is a key to the success of the programme.

Building a strong, joint integrative leadership structure to develop and manage such a programme is critical. The ongoing management is particularly key. The Terry project leader returned to full-time faculty status after the first year. The MBA director, who was managing the administrative aspects of the programme, picked up his role of managing technical infrastructure and programme content. This move created problems because the MBA director did not have the time to manage all aspects of the PwC Consulting MBA programme and the regular MBA programmes as well. Problems included such things as faculty working less as a team, technology improvements being slow or non-existent, and the relationship with PwC Consulting staff starting to weaken. To remedy this situation, the leadership of the PwC Consulting MBA was taken over by an Associate Dean and the Distributed Learning Support Group was moved under the newly created Chief Information Officer for the Terry College.

- *Develop mission/vision and communicate it*

  Implementation of this programme hinged on successful development and communication of the vision behind it. Many of the professors, staff and students had no idea what to expect from the start. It would have been very easy for them to participate in a programme taught through traditional classroom interactions, since they already knew what that entailed. Making them see that a blended delivery method would also work and work better for this context required their deep support. Such a situation requires development of a clear vision and communicating it through all means possible to garner that support and avoid failure. As outlined earlier, we created both programme and technology visions (see Figures 10.3 and 10.4) and communicated them continually to relevant stakeholders.

- *People behave in line with the way they are rewarded, thus, reward them for the behaviour you want*

  Both change and psychology literatures point out that people behave in line with the way they are rewarded. Thus, in the Terry–PwC MBA programme we wanted to make sure we rewarded people for their commitment and performance. The faculty was the key group we focused on. The faculty who teach in the programme received summer compensation, first to learn the new technology and develop their course and then to teach in the programme. Their Terry–PwC course was also given bonus credits in their regular teaching load, e.g., a

2-credit course counted as 3. All faculty in the programme received a laptop along with appropriate teaching resources they needed to teach in the programme. The recognition they received because they had been "chosen" as the best to teach in the programme is also an indirect compensation.

The students who participate in the programme are rewarded by first having the MBA paid for by PwC Consulting while still not forgoing any salary compensation. PwC Consulting structures this payment in the form of a student loan that is paid back over the three years following the programme. This structure encourages students to stay employed by PwC Consulting for five years, two during the programme and three years after. Indirect compensation was much the same as for the faculty: by being chosen by the company to participate in the programme they are being recognised as one of the top consultants. However, probably the best student reward is the MBA degree. Most students indicated that they would not have been able to get the degree without this type of programme.

- *Consider the blended approach when implementing e-Learning*

The change literature would argue that good intervention is critical to a successful change effort. Our research and experience indicate that good e-Learning intervention needs to focus on blending approaches and technology. Almost every aspect of the programme blends different approaches. The blend between distance and face-to-face education and the blending of technologies are critical keystones of the programme.

In supporting this blended approach, the faculty use a variety of tools to distribute the class materials while students are at a distance. LearningSpace is good for presenting text material or for holding other files. Some of the faculty create media presentations and put these into LearningSpace to take the place of face-to-face lectures. These approaches are very network intensive because all materials are delivered over a network. Large files that are to be used within a class would be distributed on a CD prior to the start of a course to minimise the network use and the space taken up on all the students' computers. This meant that the distribution of material had to be balanced between methods: network and CD. Currently, the Terry College is also working on online live tools for the future. The advantages and disadvantages of all the delivery methods will continue to be assessed and "remixed" to provide the most effective blended course and programme delivery.

However, the students do the most important blending. While the students are off campus, they have to balance their work life, personal life and student life. For these consultants, they are always worried about the balance between work and personal life, but with the addition of student life, they are typically one step from the breaking point.

We address this by imposing rules for hours devoted to work and study. We also address this during the on-campus visits by encouraging social activities, making sure the students know they are not alone. Teambuilding is part of this approach to combat the fatigue. While the teams are important for the learning, they are also critical for the student's emotional support. While the students are at a distance, they rely on the team to make it to the next on-campus visit. During the visit, team building/rebuilding is a key. Thus, to be effective this means that we have one more thing to balance. We have to balance team building, social events and instruction. Clearly, for this student balance to be successful, both Terry and PwC Consulting personnel have to work together to enforce rules and provide support.

- *Develop learning communities*

  Another key factor of the programme intervention was the creation of learning communities. The core community is the small student virtual learning teams (4–5 persons) through which much of the initial course-related discussion and work is done. Besides task support, these teams provide strong emotional support. As one student put it "I can't imagine doing this programme without the emotional and spiritual support of my team." We also continually do activities that develop and reinforce the class as a learning community. Students develop a network of relationships that carries over into their professional work. Students often contact each other for support on PwC Consulting projects. This networking among students was one of PwC Consulting's goals for the programme.

  The initial group of faculty plus key support staff also operated as a team or learning community. The key faculty and staff group, which had formed through the summer training prior to the first semester, helped each other through this transition process, sharing learnings and solving problems together as they emerged. Although we have not been able to maintain this community as well as we would like, the initial community of faculty and staff was critical to the successful launch of the programme. We feel the key reason for lack of sustainability was the lack of leadership in this area. This issue was discussed in the guiding coalition guideline.

  We also view all people involved in the programme as a team or community. Working together as a team, the students, faculty, and Terry and PwC Consulting programme staff have made enormous progress in establishing effective ways of teaching and learning in a blended e-Learning environment.

- *Training and support are critical for success*

  The change literature clearly documents that you should focus on unfreezing (getting the system ready for change) and refreezing (stabilising and supporting the system after the change has been implemented), and not focus on the change itself. Two of the keys to unfreezing and refreezing are training and support.

The initial core faculty cited initial and on-going support and technical and pedagogical training as a major factor in their success as programme instructors. The faculty needed to learn the technology and how to put their course into this new paradigm. The initial faculty devoted an entire first summer to learning the technology and developing their courses. As additional faculty were brought into the programme, they were trained also. Unfortunately, later groups of faculty have not been trained as well as the first group and this has caused problems.

A support structure was put together for faculty to first learn the basic technology and later was available for specific design and development questions. Once the programme started, this same support structure would train the students to use the technology and provide technology support to students and faculty during the MBA programme. Student teams are also trained in how to use the technology effectively for supporting distributed teamwork in the Team class. This approach was extremely successful. Due to changes in personnel in 2001, the support is not as good as it has been in the past. You can see the negative effects of this even though we have a well-established programme and solid technical infrastructure.

Because training and support needs were met, faculty were able to direct their energy to providing effective learning experiences for students, and students could focus on learning, not technology or other types of problems. The bottom-line is simply that training and support are critical for success especially where you are implementing new innovative technologies such as e-Learning or digital collaboration.

- *Use evolutionary approach focusing on ongoing improvement*

One of the principles of change found cited in the literature is that when there is high uncertainty in an area, you need to use an evolutionary, prototyping or continuous improvement design approach. Although we felt comfortable with the customised programme content, we knew fine tuning would be needed in courses to better meet the needs of PwC Consulting. However, from a course delivery perspective, e-Learning was a big unknown to us, especially our blended approach. Thus, our vision from the very beginning was to set in-place mechanisms to allow us to continuously improve.

This programme produced a new reality for the students and faculty involved. Implementing this programme required constant flexibility. Two key factors made this work: (1) integrated electronic evaluation and feedback, and (2) responsiveness by everyone involved, especially the faculty. Students in the programme cited the responsiveness of the faculty and administration to student and PwC Consulting concerns as a key factor for success.

Each course had an electronic feedback tool built into its structure and programme-related surveys were carried out once a week. Students commented on the problems they were facing and sent suggestions for improvements in the technology and course designs. These problems and suggestions were then grouped and sent to the appropriate place. On the individual level, faculty members received feedback on their courses. This varied from broad curricular ideas and content questions to ways to improve details of their course interfaces in LearningSpace based on what other teachers had done. This led to group sharing. Faculty members might then bring up the curriculum and technology questions/issues in regular meetings where effective solutions could be developed and implementation plans created.

We view the Terry–PwC Consulting MBA programme as an ongoing or continuous change process. An ongoing change process requires dedicated support over time to adapt both the organisation and the technology to changing conditions and technological capabilities. Opportunity-based change, changes that are not anticipated ahead of time but are implemented in response to unexpected opportunity, depends on the ability of the organisation to recognise opportunities, issues and unexpected problems when they arise (Orlikowski and Hofman, 1997). Establishing a guiding coalition/leadership and providing training and support (guidelines discussed earlier) created the foundation for continuous improvement through both anticipated and opportunity-based changes.

- *View target of change from holistic system perspective*
  The change literature presents many failures that were due to change agents not taking a holistic perspective of the change target. To change an organisation, a change agent must have some picture or target of change. In the Terry–PwC MBA implementation, a socio-technical systems (STS) view was taken. STS views the target as a set of interrelated parts: processes, technology, people and structure (Bostrom, 1983). Technology was viewed as an enabler. The focus was on how we could use technology to facilitate the learning process, the interactions between students and students and faculty, and to support team and community structures that were being created. Effective integration of processes, technology, people and structure was the outcome.

10.8 Conclusion

We started writing this chapter just before the bombings of the World Trade Centre in New York City on 11 September 2001. This event and others, along with the rapid changes in the economy in many countries dur-

ing this time, have created a very different reality in a very short period of time. Reduced budgets and the inability or unwillingness of people to travel has caused many organisations to start looking at e-Learning and digital collaboration as core mechanisms for supporting and doing business. We see this trend continuing.

At the same time, organisations continue to realise the importance of the intellectual capital of their employees. For example, many organisations in the autumn of 2001, that are implementing large layoffs and staff reductions, are faced with the sad reality that much of the intellectual capital of the leaving employees have never been captured and, thus, is lost.

The situation reminds us of the quote by Jack Welch that we used at the beginning of this chapter. He argues that organisations win by raising the intellectual capital of the organisation every day and that when people are inspired to learn, the energy and excitement generated from the learning is how you energise an organisation. PwC Consulting and the Terry College, through their joint creation of a blended e-Learning based MBA programme, have raised the intellectual capital in both organisations. The programme has also energised both organisations. For example, the learnings and the infrastructure created have helped Terry launch two additional new MBA programmes. Our efforts have made enormous impact in establishing and understanding effective ways of teaching and learning in a blended e-Learning environment. This chapter has provided a vehicle for sharing our learning to energise a larger audience.

Acknowledgements

The authors wish to thank Don Burkhard, the first PricewaterhouseCoopers Consulting project coordinator, for his efforts. Without him this programme would have never been a success! We also wish to thank all the students in the first graduating class, the Class of 2000. The inputs from this class were critical to implementing key changes that led to the success of this programme.

Part III

Conclusions and Implications

11 Integrating the Lessons Learned

Bjørn Erik Munkvold

11.1 Introduction

The six case studies in Part 2 of the book provide an in-depth view into the process and experiences of implementing a spectrum of collaboration technologies in different industrial settings. Basically, these chapters speak for themselves, offering rich contextual information, explicit findings, and advice for readers to relate to their own context.

The purpose in this chapter is to look across projects to examine similar and contrasting findings in the different cases, and identify particularly influential factors. Findings from the literature review complement the experiences from these field studies. Table 11.1 presents an overview of the cases in Part 2 to aid in this discussion.

This analysis and discussion is structured according to the taxonomy introduced in Chapter 4 (Table 4.1). The taxonomy separated four general types of factors:

1. Organisational context
2. Implementation project
3. Technology
4. Implementation process.

11.2 Organisational Context

Organisational context includes both internal and external environmental characteristics in which implementation takes place. As shown in Table 11.1, the field studies cover a variety of industries and sectors. In general, the highly contextual nature of these implementation projects makes it difficult to suggest any general findings based on variables such as type of industry, company size, etc.

Table 11.1 Overview of case studies

Companies	Sectors	Main technologies studied	Application areas
Statoil (Chapter 5)	Oil industry	Document management and workflow.	Development, storage and sharing of information.
		Meeting support technologies (EMS, application sharing, audio and video conferencing)	Supporting co-located and distributed meeting processes
Kværner (Chapter 6)	Engineering and construction, oil and gas	Global area network, document management, application sharing	Distributed engineering, project repositories, training and support
Boeing (Chapter 7)	Aerospace	Data conferencing	Support for distributed project teams
Sun, Microsoft (Chapter 8)	IT industry	Online group calendars	Meeting scheduling
Telenor R&D field trials (Chapter 9)	SMEs in building construction Public sector (telemedicine, HSE, public administration)	Integrated communications technologies, including e-mail, file sharing and document exchange, teleradiology solution, awareness support, bulletin boards, discussion forums	Electronic tendering and distributed projects, teleradiology, awareness support, discussion and experience transfer, communication and information sharing
University of Georgia/ Pricewaterhouse-Coopers Consulting (Chapter 10)	University, Consulting	e-Learning technologies: Learning Management System, audio/video-conferencing, instant messaging, discussion databases, document sharing	Support for distributed MBA programme based on virtual learning teams

Except for the organisations included in the field trials described by Telenor R&D, all organisations are large companies operating on an international scale. The main contextual difference that can be identified relating to company size is the internal IT resources available, in the form of infrastructure, manpower, experience and competence. However, despite access to more internal IT resources the large companies have also experienced problems and delays in their implementation projects, as a result of the complexity and heterogeneous nature of these projects.

Most of these companies operate in highly competitive environments. The exceptions here are the organisations in the public sector served by Telenor R&D. The academic sector, here represented by the University of Georgia, also faces increasing competition from the growing number of educational programmes. Thus, the development of the PwC Consulting e-Learning programme can also be seen as a response to a competitive market. Finally, Statoil is in a somewhat "sheltered" position, being a partly state-owned company and the major Norwegian oil company. Still, this company is also highly dependent on market conjunctures and fluctuating oil prices.

11.2.1 Internal IT/IS Organisation

The role and function of the internal IT/IS organisations varies to a large extent among these cases. Statoil and Boeing both have large, centralised IT/IS units, responsible for maintaining the IT infrastructure in these companies. Statoil IT is also responsible for identifying emerging technologies and adapting these as internal services in the company. There is a dedicated unit for maintaining all activities related to collaboration technologies (the e-collaboration team). In Boeing, this responsibility is maintained by a separate unit in the applied research and development organisation. This unit then recommends new technologies to be included in the portfolio served by the IS organisation in Boeing.

Kværner is currently downsizing its centralised IT function, changing its role from a large, independent unit to smaller, localised service providers for the different engineering disciplines. Maintenance and further strategic development of their global area network is outsourced to an external telecommunications provider.

The implementation projects in Sun and Microsoft both include collaboration tools developed internally. This implies a high level of internal IT competence related to these services.

The organisations involved in the field trials reported by Telenor R&D represent a different world, in that these small companies and public organisations hardly have any separate IT/IS function at all. The responsibility for maintaining IT resources lies with one or a few persons, often as part of another job. The general problem is therefore limited capacity that is consumed by handling the day-to-day pressing matters. Thus, technology implementation in these companies is dependent on an external partner such as a vendor or a research group like Telenor R&D.

In the Terry College–PricewaterhouseCoopers Consulting programme a dedicated technical support organisation, involving three full-time employees and nine part-time assistants, was established for managing all aspects related to the e-Learning infrastructure. The university also saw the project as an important means to build the competence and infrastructure that would enable offering similar e-Learning programmes to other companies.

11.2.2 Collaborative Culture

The literature review identified three important elements of the internal organisational context:

1. the degree of collaborative work practices in the adopting organisation;
2. the users' felt need for technology supporting this collaboration; and
3. the organisation's reward systems.

Together, this is often seen to constitute the "collaborative culture" of the organisation. The literature argues that the existence of such a culture is vital for the implementation of collaboration technology.

How does this relate to the case organisations studied in this book? For many of these organisations, the implementation projects studied represent their first encounter with collaboration technology, at least for supporting the particular work process. To be able to address the question of existing collaborative practices and the users' felt need, it is necessary to analyse (a) the origin of these projects and (b) the form of work practices that were replaced in the companies. Why were the projects initiated in the first place, and by whom? Was this driven by actual business needs or simply "because everyone else was doing it"? What work practices were replaced by the technology implemented? Table 11.2 provides a brief summary of these questions for the organisations studied.

Project Origin

Table 11.2 shows that many of these projects have been technology-driven rather than based on a perceived need expressed by users. The possibilities represented by the emerging technology have created new opportunities for structuring and conducting work. In several of these organisations, this potential was identified by organisational units dedicated to the investigation of emerging IT technologies such as collaboration technology.

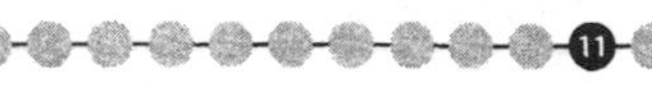

Table 11.2 Project origin and existing work practices

Organisation	Project initiative/origin	Work practices replaced
Statoil	Lotus Notes: middle managers being introduced to the technology	Case handling using former e-mail and archive system
	Desktop conferencing (Intel ProShare, NetMeeting): person responsible for PC hardware in Statoil IT Electronic meeting systems (GroupSystems): Lotus Notes project leader from Statoil IT	Traditional co-located meetings
Kværner	KIGAN/KINET implementation: central IT staff in the Kværner concern management	Limited prior collaboration between different companies in the Kværner group
Boeing	Exploration of data conferencing technologies conducted by a dedicated unit (Mathematics and Computing Technology) within Boeing's applied R&D organisation, Boeing Phantom Works	Co-located project meetings imposing extensive travel for many employees
Sun, Microsoft	Internally developed product – bottom-up adoption spreading from individual developers to managers and administrative assistants	"Manual" meeting scheduling
Telenor R&D field trials	• Virtual construction company: network coordinator employed by SMEs • Teleradiology project: Telenor R&D as part of larger activity on telemedicine • Local politicians' channel: Telenor R&D • HSE Net: Telenor R&D	• Project communication through phone and fax • Radiologist travelling between hospitals • Communication and information exchange using paper, fax and phone
University of Georgia/PwC Consulting	Prospect for specialised e-Learning programme developed by MIS faculty at Terry College of Business, University of Georgia	Traditional on-campus MBA programmes

In Statoil, Statoil IT is responsible for identifying and translating the potential of new technologies into internal products, based on their knowledge of the business operations and potential needs. They then make these technologies available and market them within the organisation. As described in Chapter 5, a formal product development process is established for this in Statoil IT. Still, as illustrated in Table 11.2, the actual initiative often comes from individuals based on their special interest and competence in an area. It is here also interesting to note how the GroupSystems implementation was initiated on the basis of experiences with the limitations of another collaboration technology (Lotus Notes) for supporting meeting processes.

In Boeing, a dedicated unit within their applied research and development organisation has responsibility to identify the new technologies. This unit then needs to "sell" the new technology to the central IS organisation in charge of the IT infrastructure. (Some challenges in this process are discussed related to the implementation process later in this chapter). Both Telenor R&D and University of Georgia have served the role as technology providers in their projects (although through third-party vendors), identifying potential uses of the technology and being responsible for transferring this to the adopting organisation or collaborative arrangement. Also in Kværner, the initiative for the implementation of KIGAN came from the central IT staff.

The implementation of online calendar systems in Sun and Microsoft was also technology-driven, in the sense that these companies were using their own products. However, the adoption of these systems was more bottom-up, spreading from individual developers and administrative assistants.

Thus, only in a few of these cases has this technology-driven process been based on explicitly stated needs for technology support from the employees. As collaboration technology often represents new functionality and new ways of working, users cannot be expected to be able to identify and express their needs for this type of technology support, even if there are aspects of the current work process that they are not satisfied with. Thus, as exemplified in several of the cases studied, this lack of explicitly felt need among the users may actually constitute an initial barrier in the implementation process. The implementation team needs to convince the users of the potential benefits offered by the new technology.

Previous Work Practices

As listed in Table 11.2, most of the work practices that have been changed or replaced by the new technology have been of a collaborative nature. The technology has mainly contributed to making these more efficient.

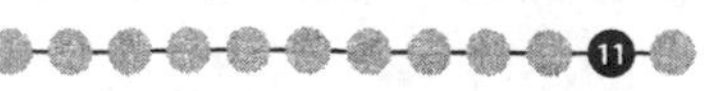

However, does this mean that a "collaborative culture" existed prior to these implementations?

Previous research (such as Orlikowski, 1992) discusses the importance of the organisation's policy and reward systems for stimulating collaboration. In none of these cases is there any particular incentive system in use for stimulating collaboration. None of the studies in the literature review report this either. Thus, while this is often recommended as important for stimulating collaboration, few organisations seem to have taken this step.

Another indicator for collaboration in the organisation is the degree of individualistic versus collective work. With most work organised as projects, work in project teams is widespread in these organisations. In general, the implementation of this technology is also intended to increase the level of collaboration in the organisation. However, the pressure from external competition often results in limited focus, constraining the possibility for sharing information and contributing to shared repositories and thus acting against the goal of increased collaboration.

There are also examples of how existing collaborative processes can act as a barrier for the implementation of new technology. In Kværner, the existing collaboration with suppliers in the local market engendered some reluctance towards adopting a new technology intended to support distributed collaboration with "remote", competing suppliers within the Kværner group.

To what extent, then, can the culture in these companies be described as collaborative? As discussed by Karsten (1999) this is a complex question. In most organisations there is no one, "distinct" culture, but rather several "sub cultures" of a more dynamic/transient nature. This is especially complex in large, global organisations (conglomerates) such as Kværner. As described in this case, there exists no "Kværner culture" as Kværner actually consists of a large number of "independent" companies that have been acquired through mergers and fusions, etc. As such, the case studies presented here do not allow for any general conclusions about the "collaborative culture" of these organisations.

Two final observations can be made regarding the importance of a collaborative culture in the implementation of collaboration technologies. First, the existence of collaborative work processes in the organisation should not be regarded as a prerequisite for successful implementation of collaboration technology. Despite varying levels of former collaborative work practices in the organisations studied, adoption and use of the collaboration technologies has finally taken place, although at a somewhat varying pace. Second, in cases such as global virtual teams where collaboration technologies represent the only viable means of conducting teamwork, this perceived need can effectively motivate the adoption and effective use of technology, even if a pre-existing collaborative culture is lacking.

11.2.3 External factors

The field studies reviewed in Chapter 3 mostly focused on implementation factors internal to the organisation. The case chapters in Part 2 also illustrate how factors related to the external environment have been highly influential on the implementation projects.

Economic Conjunctures

As with all technology investments, the market situation and economic conjunctures may have a high impact on the resource situation for this type of project. There are examples of how cost reduction has both stimulated and hindered the adoption of collaboration technologies. When Telenor R&D was implementing Lotus Notes for supporting the product development process, this project was among the first activities to be stopped when budgets had to be cut. As discussed earlier, the intangible benefits from some collaboration technologies make them "fragile" also in the sense that they risk a sudden death when numbers go red. However, the opposite reaction occurred at Statoil when low oil prices created the need to reduce travel costs. This in turn gave momentum to an organisation-wide campaign for using video and data conferencing to support distributed meetings. When the pressure to cut costs eased, use declined somewhat.

In inter-organisational collaboration, the situation in local markets may also be important for the organisations collaborating. For example, in the Virtual Company project run by Telenor R&D in the construction industry, the local market conjunctures were decisive for the priority given to the collaborative arrangement. When the local market was slow, increasing emphasis was given on the virtual collaboration with other companies to develop new possibilities for entering other markets. However, as soon as the demand in the local market increased again, individual projects regained full priority at the expense of the network collaboration.

In general, this type of factor is difficult to plan for and impossible to control. Thus, the important thing for an organisation here is to build sufficient flexibility and "improvisation capability" into the process, to be able to respond to such changes and try to identify new possibilities as a result of the changing conditions (cf. the change model of Orlikowski and Hofman, 1997).

Relations to Third Parties

There are several examples of how relationships with different third parties have greatly influenced implementation and use of collaboration

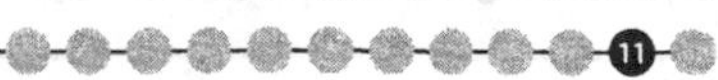

technology. These parties may be customers, suppliers, vendors or the local community in general. For example, in Kværner Australia, the heterogeneity in projects regarding scope and size combined with customers' preference for co-located project operations, represented a barrier to applying what Kværner would define as best practice, such as their "high-value, low-cost engineering" concept based on distributed project operations.

The importance of local markets in distributed organisations was mentioned in the previous section as well. In Kværner, the existing network and collaboration with the local market and community initially comprised a barrier to the vision of increasing distributed collaboration among the Kværner companies. Similarly, in one of the SME networks participating in the Virtual Company project in Telenor R&D, customers expressed concern when the SMEs together formed an alliance that would potentially become a competitor in the same market as these customers. Afraid of damaging their relations with these customers, some companies were reluctant to invest resources in this network collaboration and related collaboration technology.

Existing ties with vendors and their technology platforms may also represent a barrier to deployment and adoption of new collaboration product suites. For example, in organisations using Microsoft products for general office support, additional implementation of Lotus Notes for collaborative applications represents a "political" decision, implying a need for extra resources and competence to support an additional platform.

The Boeing case also describes how the vendors' lack of focus on technology integration was a barrier in their deployment of data conferencing technology. In general, companies wanting to acquire collaboration technology are faced with the challenge of having to relate to several vendors, each promoting their own product suites and concepts for collaborative support. For larger organisations representing significant customers, one proactive strategy is to challenge the vendors by doing a systematic, comparative evaluation of these concepts. An example of this is the evaluation made by Statoil prior to selecting the technologies for their next generation collaborative infrastructure. In doing this, Statoil was also able to actually influence the vendors' solutions and strategies related to the products.

A final challenge related to collaboration with third parties is described in the Kværner case. This involved the legal aspects of information ownership and accountability. As part of their B2B strategy, Kværner makes increasing use of shared project archives and information portals where engineering data and project documentation are shared with clients, subcontractors and vendors. The problem is who maintains the copyright to this documentation as it evolves during and after the project. Further, potential liability from electronic documentation as court evidence in conflict situations has somewhat restricted the possibility for making the "project close up reports" a standard practice, and putting these on the company intranet.

Global Issues

In global spanning projects such as the KINET implementation in Kværner, varying practices at the national level may also impact implementation. For example, protective policies enacted by national telecommunications providers constituted a barrier for the implementation team when establishing operations in the Asian region. In addition, varying levels of sophistication of IT infrastructure in the different world regions posed challenges to developing a global network topology serving all offices.

National culture is also an important area with regard to implementation of collaboration technologies across national boundaries. Although not addressed explicitly in the case studies presented in this book, the importance of understanding the local culture in which the implementation takes place is widely acknowledged. Culture exerts important influences on decision-processes and user perceptions of new technology. In general, it is important to remember that the context of all the companies studied in this book is the "developed" world. The experiences cannot automatically be transferred to organisations in other regions of the world.

Finally, major global issues such as terrorism and regional conflicts will of course also influence the general situation. For example, fear of travelling and economic downturns following the September 11th bombings have increased the demand for videoconferencing equipment supporting distributed meetings.

11.3 Implementation Project

As can be expected, the organisation and conduct of the implementation project is of key importance to its outcomes. Both the literature review and the case studies present lessons related to this. These lessons are organised under six areas including implementation strategy, implementation team, pilot projects, user involvement, training and supporting roles.

11.3.1 Formalised Implementation Strategy

Most literature on technology implementation is clear on the need for a well-planned approach based on a formalised implementation strategy.

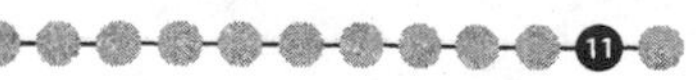

However, both in the literature review and the case studies in Part 2, there are several examples where this was only fulfilled to a limited degree. For example, in the heterogeneous and complex organisational setting of the Kværner group, the KINET implementation was more of an ad hoc nature. In hindsight, the project co-ordinator expressed how using a more planned approach could have saved them from some of the problems in the project. Still, both the development of technical services and the marketing and diffusion of these in many ways were a learning experience for the implementation team as well as the vendor. In contrast, the Terry–PwC Consulting e-Learning programme applied a formalised approach from the outset. Building upon the project and technological experience of the key persons involved, programme and technology mission/vision statements were defined during the pre-planning stage. In general, the delineated focus of this project regarding scope of the technology implementation supported this structured approach.

Two exemplar cases of well-planned projects with adequate resources are the implementation of Lotus Notes in Statoil and the implementation of data conferencing in Boeing. The internal IT resources in these organisations clearly contributed to the relatively successful rollout and deployment of these technologies. In both of these organisations, however, the formalised implementation projects emerged only after several years of technology exploration.

Also, as discussed by Palen and Grudin in Chapter 8, successful deployment of the technology does not guarantee adoption by the users. In the Statoil case, we also see how the actual user adoption of different Notes applications varied. In general, the product development process in Statoil IT is intended to ensure a standardised approach for IT implementation projects in Statoil. Still, the implementation of GroupSystems in this company was of a more evolutionary nature than the Lotus Notes implementation, even though this was also conducted within the framing of the product development process. This illustrates how the nature of the implementation of different collaboration technologies may vary, even within the same organisation. This will be discussed further in the section related to the implementation process.

The implementation projects described by Telenor R&D were conducted in a research context, although Telenor R&D also acted as technology provider. In many ways this was a challenging role for the researchers in that the customers in these projects, represented by the participating organisations, expected Telenor R&D to offer similar support as a "commercial" vendor, whereas they mainly wanted to handle these as field trials.

The implementation of online calendars in Sun and Microsoft represents a special case where the technology deployed was developed internally. After being made available on the company network, the technology diffused through "natural dissemination" rather than managerial mandate, based on social influence mechanisms such as peer pressure. This is further discussed related to the implementation process.

11.3.2 Organisation of the Implementation Projects

The most common organisation of the implementation effort is in the form of a project, led by an implementation team. The implementation teams responsible for deployment of the technology are often very small compared to the scale of the implementation project. For example, the KINET implementation in Kværner serving installations in four world regions, was led by a core team of only five people. Similarly, the implementation team of a Lotus Notes application in ABB Corporate Research, serving 800 users in eight different locations, only consisted of three persons (Munkvold, 1999).

The main benefits of this small implementation organisation are flexibility, ease of coordination and the ability to make decisions and act on them quickly. The downsides of such a structure are the occasional extreme pressure and workload on the team members, and the resulting vulnerability of the project from being dependent on a few key persons. For example, in Kværner this led to capacity problems generating a backlog of activities.

These teams typically include a mix of business, project and technological competence and experience. In Kværner, problems with finding internal staff with this competence mix as well as the reluctance to hire external people were also used as arguments for keeping the implementation team small. In some of the organisations (Statoil, Telenor R&D and Microsoft), this multidisciplinarity is extended to also include psychologists, sociologists and anthropologists. While the integration of these different backgrounds in the same team can sometimes be demanding, it may provide an invaluable resource in obtaining a more holistic perspective on the implementation process.

In the Terry–PwC Consulting MBA project, the implementation was conducted by a dedicated Distributed Learning Group, involving three full-time employees and nine part-time assistants. Also, there was close collaboration between the Terry project leader and the PwC Consulting coordinator. In Boeing a permanent unit within the applied research and development organisation conducted the initial investigation and prototyping stages. Later, the responsibility for the technology was transferred to a dedicated support environment, including a product manager and extensive training and technical support.

The Statoil case is an example of how the organisation of the implementation of collaboration technologies may evolve over time. From originally being organised in smaller teams dedicated to specific collaboration technologies (Lotus Notes, data and videoconferencing, meeting

support systems), there has been a gradual merger of these different teams. The integrated e-collaboration team represents the final stage in this process, being responsible for adaptation, deployment and support for all collaboration technologies in the organisation. However, as this team is staffed with persons from the prior dedicated teams, there is still a way to go before the team becomes fully integrated with respect to technological perspectives and priorities.

11.3.3 Pilot Projects

The literature widely recommends using pilot projects for this type of implementation. In several of the cases discussed in the book, this has been used as an important basis for initial learning about the possibilities and limitations of the technologies, as well as for identifying potential benefits for further internal marketing/sale of the technology.

Several studies in the literature point to how the composition of pilot groups is vital for a successful pilot project. If this group does not reflect a real user group or use situation, it cannot be expected that the pilot will effectively document the benefits of the technology. This was also experienced in one of the virtual organisations served by Telenor R&D. Here, the different companies participating in the project appointed their own "representatives" to participate in the pilot group. As it turned out, different criteria were applied in this selection process, such as IT competence/experience or job function in the collaborative arrangement. The result was a pilot group consisting of members with no real need to communicate, and in consequence the technology was hardly used. This illustrates how the implementation team should be involved in the selection of pilot users, especially for projects involving distributed and/or interorganisational collaboration.

11.3.4 User Involvement

The involvement of users in implementation projects is frequently emphasised as an important condition for success. In small organisations, the scope of the implementation project often makes direct user involvement possible. However, this can also be restricted by the lack of IT competence among the users and/or lack of time to participate. For example, in the Telenor R&D field trials involving SMEs and the public sector, the researchers needed to guide the users through the entire process, from identifying and specifying requirements to establishing routines for using the technology. Although users were represented in design meetings and took part in prioritising among the different needs, the researchers in fact made the main design decisions based on their understanding of the case.

In large heterogeneous companies, one of the aims of implementing collaboration technology is often to support some form of standardisation related to communication and information sharing. This was the case in both Statoil and Kværner. In these projects, ensuring representative user involvement in the design of the new solutions is often not considered a major issue, as the focus is more on establishing new "best practice". However, this approach involves the risk of not capturing important "local" needs of different units in the company, that may be vital for successful adoption and use of the technology.

11.3.5 Training

Previous studies strongly emphasise the importance of training in the implementation projects. In many cases, training is reported to have been inadequate, only focusing on the "mechanical" functions of the technology and not on how to relate the functionality of the technology to the work tasks needed to support collaboration. This resulted in ineffective use of the technology, and not realising the full potential of this.

In the case projects in Part 2, training was emphasised to various degrees and delivered in different ways, including combinations of training courses, individual training and web-based training material. The most extensive training reported was related to the implementation of Lotus Notes in Statoil. Still, although this training also included a focus on routines for local adaptation and use of the technology, the process of actually establishing this use was long. This illustrates that training needs to be followed up by incentives to adopt the new work practices and some form of audit to ensure that these practices are continued.

Tight budgets and competition are clearly barriers to investing time and money in extensive training. For example, Kværner often does not have budgets for training temporary personnel to use their proprietary engineering applications. Thus, many of the temporary engineers had to use valuable project time to learn these tools. Kværner therefore tries to convince their customers that the project will benefit from including this training, and gradually experience greater acceptance for this among their customers.

For team-based collaboration, providing training in teambuilding is also important, especially for virtual/distributed teams with limited face-to-face interaction. The importance of including some "face time" together in the initial stage of the project for introduction and building social relations is emphasised in both the literature and the case projects. For example, in Kværner they usually start major projects with face-to-face meetings involving different teambuilding activities. Later in the projects, videoconferencing is also used as an important medium for supporting the development of interpersonal relationships in the

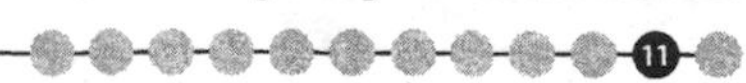

distributed engineering teams. Also in the e-Learning context of the Terry–PwC Consulting MBA case, the initial two-week seminar was reported to be crucial to the success of the programme, involving training of students in technology and also team skills.

11.3.6 Supporting Roles

The case studies have highlighted several supporting roles, beyond the formal project organisation chart. In the following these roles are briefly discussed as related to the implementation projects.

Champions and Change Agents

Chapter 4 discussed the importance of a project champion as expressed in previous implementation research. This is also underlined by the projects presented in this book. However, the function and formal position of this champion role has varied. In some cases, these champions have also played the role of change agents (Ginzberg, 1978) and "missionaries" for the technology.

In Statoil, the implementation of meeting support technologies and the later integration of these (NetMeeting, GroupSystems, audio and videoconferencing) is a clear example of an implementation project that was almost totally dependent on a single champion. Without this person's persistence and faith in this technology, GroupSystems would most likely never have been implemented in Statoil. As a senior engineer in Statoil IT, this person's role as a project champion was based on his competence and vision of the potential of this technology rather than any formal, managerial position.

Similarly, in Boeing, the group within the applied research and development organisation responsible for technology exploration served as champions in promoting the technology in the organisation and working actively for diffusing the technology throughout the company. This involved important functions such as enrolling management as "allies" by exposing them to the technology through demonstrations of its potential. Further, this group also had to "sell" the technology to the IS organisation in Boeing, and convince them to include this in the company's IS infrastructure and establish the necessary support environment. This involved the assignment of product managers for each of the key technology components, responsible for supporting and evolving this service. These responsibilities involve communication with vendors, evaluation of new products and product versions for potential replacement or upgrade of existing technology, and customisation of the products. Further, the product manager is also responsible for developing websites for training and offering procedures and guidelines for use. Due to this proactive role of the product managers in leading the development of the services, they are referred to as change agents in the case study. The support environment in Boeing also includes a hotline offering 24-7 technical support.

In Kværner, the project coordinator for the KIGAN/KINET implementation played the role of a champion in the early stages. He held the overall vision of what this global network infrastructure could mean for Kværner in terms of making communication more efficient and also increasing collaboration, and was responsible for the internal marketing of this concept and the technology solution. This involved visiting the different companies in Kværner and convincing management there to adopt the technology.

In the Telenor R&D cases, the whole team of researchers acted as champions through their enthusiasm and engagement in the project. This also involved customer support. In addition, employees in the "target organisations" in these field trails played important roles as local champions. In the virtual construction company the network coordinator worked together with the research team in deploying the communications solution. As coordinator, he was the only person with an overall view of the activities in the network and the related need for communications services. Similarly, in the case of the local municipality, the mayor served as the champion and stimulated adoption and use of the technology among the politicians. In all of these cases, the champions also were in a formal position to act on their suggestions.

In the Terry–PricewaterhouseCoopers Consulting project, the Terry faculty member serving as project leader and the PwC Consulting coordinator acted as change champions. Although this project faced less "resistance" and inertia within the organisations adopting this new programme, the teamed skills of these champions and their ability to mobilise resources in their organisations were key to developing the programme. Later, the guiding coalition expanded to include the core group of faculty selected for the programme.

Interestingly, only in the case of calendar adoption in Sun and Microsoft was the presence of a champion not emphasised by the authors. Instead, social influence mechanisms such as peer pressure from colleagues were found to influence the adoption pattern more.

Superusers and Mediators

The appointment of so-called superusers is a frequently used strategy for supporting adoption and use of a new technology. These users may be given special training for taking on this role, or come to fill this function through their natural skills and interest in the system. Thus, the existence of superusers is not necessarily based on a planned approach. These users constitute a local support function for other users in their working environment, helping them with problems related to "technical use" of the product as well as mentoring related to effective use to support specific activities. Often, these superusers also serve as the contact point for the IS organisation.

A related term discussed in the literature is mediators, denoting users that contribute to the implementation process by adapting the technology to the local context and shaping user reaction with it (Okamura et al., 1994). The role played by the network coordinator in the virtual construction group and the mayor in the local municipality in the Telenor R&D field trials clearly has elements of such mediation. Also, the local IT coordinators in the different Statoil units played an important role in adapting the technology to local needs.

Facilitators

Although collaboration technology is gaining ground and has reached a level of maturity, problems with realising the expected benefits from different collaborative technologies in their daily use are still reported. Several therefore see a need for dedicated experts that can facilitate effective use of the technologies and disseminate best practices. The importance of meeting facilitators has long been acknowledged, both for "traditional" and electronic meetings. The Boeing and Statoil case studies discuss how the virtual/distributed meetings require facilitation to ensure quality group communications and effective appropriation of the technology. For example, for data conferencing in Boeing, having dedicated personnel prepare the virtual meetings was important to avoid the usual technical delays.

As discussed in this section, some of these supporting roles (champion, superuser, mediator, facilitator) may not necessarily be a result of formal appointment as part of the implementation strategy, but rather emerge naturally throughout the implementation process. In either case, it is important that these roles are acknowledged and stimulated by allocating sufficient time and resources for the personnel filling these roles.

11.4 Technology-related Issues

The taxonomy in Table 4.2 divides technology-related factors into general and specific factors. In the following, these two categories of factors are summarised related to the findings from the implementation projects in Part 2 of the book.

11.4.1 General Factors

A common view related to the organisational implementation of any type of IT is that the main challenges are organisational and not technological per se. While this often may be a valid observation, there is a potential danger here of underestimating the importance of the technology. As seen in all of the implementation projects in Part 2 of this book, technological issues may actually have a significant bearing upon such implementation projects. Clearly, without a well-functioning technological infrastructure, the potential benefits and organisational effects will never occur and you never reach the stage where organisational change becomes important! Although technical problems are often more easily identified and addressed than organisational problems, the resulting project delays and user frustration may lead to distrust in the technology and thus harm the adoption process. (See also Chapter 9 for a discussion of this argument).

Various forms of technological incompatibility have been a problem in several of the cases. Examples of this are incompatibility between new and existing e-mail systems in both Statoil and Kværner, and between different LANs in the virtual construction company. In Kværner, the problems related to developing a flexible network topology created major delays in the KINET project. Stability and performance of the new technology has also been a problem related to several of the new services. Many of these technologies still were relatively immature at the time of installation.

There are also examples of how the collaborative features of a technology may not be fully utilised due to restrictions on the basis of performance or security concerns. For example, in Boeing the IS organisation's concern about network impact from using NetMeeting resulted in prohibiting audio and video conferencing as part of the NetMeeting service. Also, security concerns related to Internet access to NetMeeting resulted in the requirement that all meetings had to be hidden and require passwords, which in turn caused problems with usability. Another example is from the early stages of Lotus Notes implementation in Statoil. The security setup in Notes enables specifying detailed access rights for different user groups. Initially, access rights were defined according to the formal organisational structure; they soon found that this hindered all the lateral collaboration between the different units in the organisation.

Usability was also an important issue in some of the cases. In the implementation of NetMeeting in Boeing and GroupSystems in Statoil, modifications to the user interface for these products were necessary to improve usability related to log-in and initialisation of the communication/meeting sessions. This illustrates that these technologies still have some way to go before the user threshold is equally low as for other communications media (telephone, fax, etc.).

The balance between perceived benefit and extra workload from adoption of a new technology is another frequently cited implementation factor related to collaboration technologies. In several organisations with distributed operations, replacing established work practices with new practices based on collaboration technology was initially perceived to imply extra work without any obvious benefits for the employees. This included the replacement of existing applications with Lotus Notes-based tools in Statoil, and the KINET implementation in Kværner. The initial reluctance among many Kværner companies towards adopting this technology also illustrates how perceived disparity in work and benefit may not only affect technology adoption at the individual level, but may also impact on the process of establishing a critical mass of adopting organisational units.

The Statoil case also illustrates how the adoption processes for several collaboration technologies may be inter-related. Internal competence related to these technologies is often a scarce resource, and parallel implementation projects for different collaboration technologies may thus compete for the same resources. Also, the preference for a specific product within some stakeholder group may sometimes reach a more "ideological level" thus resulting in internal competition of a political nature. In all, this calls for an integrated perspective on the various activities related to the collaboration technologies.

11.4.2 Technology-specific Factors

An important argument made in this book is that when implementing a collaboration technology one needs to focus explicitly on the specific characteristics of this technology. Much of the previous practitioner-oriented literature on implementation of collaboration technology has tended to be rather general. All collaboration technology gets lumped into a single, unified concept ("groupware") without addressing how the special characteristics of each technology may influence the implementation project. This book therefore attempts to present a more detailed analysis of the implementation related to various collaboration technologies. Table 11.3 presents an overview of implementation experiences from the cases in Part 2, related to each of the five categories of technologies defined in the framework in Chapter 2 (Table 2.1).

The table illustrates how various aspects related to infrastructure, functionality, usability and routines have been important for the different technologies. For communications technologies, technical compatibility and relative advantage have been key concerns. Data conferencing is a relatively new technology in use, still characterised by some technical challenges and restricted use of the application sharing functionality. Related to document repositories, a key issue has been to establish

Table 11.3 Specific implementation experiences related to each case

Technology	Cases	Implementation experiences
Communication technologies		
E-mail	Statoil, Kværner	Problems related to integrating different e-mail systems in the organisation.
	Telenor R&D trial in SME virtual company	Lack of access to all participants in the network resulted in early adopters discontinuing use.
Videoconferencing	Kværner	Important for teambuilding in early stages of project.
	Boeing	Limited use – not considered to offer much extra value beyond audio and data conferencing.
Shared information space technologies		
Data conferencing/ application sharing	Boeing	Increasing use for supporting virtual project meetings. Interface modifications needed to improve usability. Concerns related to network impact and security resulted in restricted functionality. Need for distributed facilitators to reduce technical problems and delays, and for improving communication.
	Kværner	Frequent problems and delays with setting up meetings. Good experiences from use of application sharing for training and troubleshooting. Less effective for distributing presentations due to bandwidth limitations.
Document repositories	Statoil	Slow adoption, lack of incentives for contributing to shared databases, "fear of exposure" among users, use gradually increasing.
	Kværner	Project archives and engineering data repositories key to distributed project operations. Also used for external collaboration. Some problems related to timely registration of data in the engineering databases, as the engineers did not see the need for this. Also problems related to lack of training of temporary workforce in Kværner's proprietary systems.
Electronic meeting systems		
GroupSystems	Statoil	Evolutionary process with several barriers: access to trained facilitators, lack of management attention and support, difficult to diffuse the technology outside Statoil IT. Some internal competition for resources with "rival" technologies. Project dependent on single champion.

(cont'd)

Table 11.3 Specific implementation experiences related to each case (*cont'd*)

Technology	Cases	Implementation experiences
Coordination technologies		
Workflow	Not explicitly addressed in these case studies. See Table 4.1 for factors identified in literature review.	
Online group calendars	Sun and Microsoft	Widespread use of online calendars compared to similar study ten years earlier. Bottom-up adoption supported by peer pressure. Improved functionality, user interface, and network infrastructure were contributing factors. Use practices varied in the two organisations regarding openness and privacy settings.
Integrated products		
Collaborative product suites (Lotus Notes)	Statoil	Varying adoption patterns for different applications. Rapid adoption for basic support tools (e-mail, group calendar), slower adoption of shared information databases (see above on "Document repositories"). Increasing emphasis on web integration.
e-Learning technologies - LearningSpace - NetMeeting - instant messaging (ICQ) - discussion db and document sharing (TeamRoom)	Terry College/ Pricewaterhouse-Coopers Consulting	Main challenges related to developing ways for effectively balancing classroom and e-Learning environments. Examples of issues were management of student workload and how to best distribute class materials. LMS offered limited support for management at programme and organisational level. Selection of technologies was partly based on fit with existing infrastructure in PwC Consulting. Close collaboration with university and client key to programme success.

incentives and routines for effective use, including data input. The experiences with EMS in Statoil illustrate how this technology requires an extended infrastructure, including electronic meting rooms and trained meeting facilitators. Also, this case illustrates how different collaboration technologies may actually end up as "rivals" in competing for scarce resources and key personnel.

The studies of online calendars in Sun and Microsoft provide an example of how a relatively "small" technology with low visibility in the organisation, has become an integral part of everyday work practices of the employees. The process by which this has occurred was mainly bottom-up, supported by peer pressure. For integrated technologies such as Lotus Notes and e-Learning systems, the main challenge has been to develop best practices related to the use of the different applications and services offered in these systems.

Again, this should not be seen as an attempt to generalise findings for the different technologies, but to attune the reader to the fact that key implementation factors may vary for each type of technology.

11.5 Implementation Process

The formal strategy and organisation of the implementation projects provide the basis and framing for the implementation process. However, as discussed earlier in this chapter, a range of contextual and technology-related factors influence the actual nature and characteristics of this process.

The case studies in Part 2 provide an in-depth description of the process related to adoption and diffusion of the technologies in the organisations. Together, these studies illustrate how the nature of this implementation process may vary for different technologies and different organisational contexts. The Statoil case also shows how different technologies within the same organisational context may follow different "implementation trajectories" (cf. Figure 5.1).

In most cases, the implementation process has spanned not months, but several years. Data conferencing at Boeing, for example, took more than seven years from the first exploration of research prototypes until NetMeeting was established as a service. In Statoil, implementing GroupSystems and desktop conferencing took two and three years, respectively.

More often than not, the implementation has taken longer than expected at the outset, due to various barriers encountered in the process. Clearly, this depends upon technological factors such as the maturity of the technology regarding stability, performance and compatibility. However, the barriers encountered are also often related to unforeseen conflicts between organisational context and technology characteristics.

Furthermore, the organisations need time to build experience with the different technologies. In general, it is difficult to define an exact time frame for these projects, as they are of a more continuous nature. For example, ten years after the implementation of Lotus Notes in Statoil the use of this technology continues to evolve. This supports the notion of collaboration technology as drifting technologies, requiring improvisation, learning and maturation (Ciborra, 1996). Due to the potential of collaboration technology to enable significant changes in organisational work practices and decision processes, the adoption of these technologies often becomes a political issue.

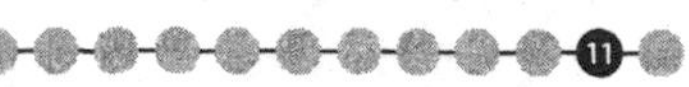

The type of technology also clearly influences the adoption pattern. While some services relatively quickly attract a critical mass of users, others require a longer adoption period involving extensive marketing. An example of the former would be Lotus Notes, with many organisations experiencing a high demand for this technology, sometimes leading to a hastened rollout with insufficient capacity for support and training. However, even for Lotus Notes, while some of the basic services in the product (e-mail, online calendar) are easily adopted, applications that require more extensive user contribution (shared databases, document archives) may take longer to adopt. An example of the second category of technologies is electronic meeting systems, as illustrated by the long and evolutionary adoption process in Statoil.

Another barrier that could slow the adoption of collaboration technology is that it is often difficult to place the direct "ownership" of this technology. At Boeing, data conferencing technology was transferred from a separate R&D unit to the central IS organisation. While many users could benefit from these services, which organisational unit should actually be responsible for the main costs? Also, as expressed earlier, it may take time before a new collaborative tool has become adopted and placed in regular use. Thus, in the infant stages, without clear organisational ownership and user demand, this technology is far more at risk of losing its priority compared to "standard" IT applications with clear constituencies in the organisation (office support, accounting, payroll, etc.).

Not surprisingly, the most effective driver for adoption is when the users feel a strong need for the technology. There are several examples of how events and changes in the organisation at a later stage have created this need for technology, thus spurring the adoption process. One example is Boeing's merger with Rockwell and McDonnell Douglas, that led to an increased demand for distributed collaboration and virtual meetings. Similarly, in Statoil the focus on reducing travel costs stimulated the rapid diffusion of MS Netmeeting.

Compared with the other cases studied, the technology implementation related to the Terry–PwC Consulting e-Learning programme went relatively "smoothly", without major barriers in the process. Since the very concept of this programme rests on the foundation of the e-Learning infrastructure, all stakeholders were equally motivated for adoption. In general, in cases where adoption and use of the technology is a prerequisite for the entire collaborative arrangement, the question of adoption versus non-adoption becomes largely irrelevant. Still, the process of establishing routines for effective use of the technology may be of a more evolutionary nature. For example, after the technology installation and start-up of the Terry–PwC MBA programme, the faculty engaged in a mutual learning phase of finding the best way of delivering the course contents based on the "blend" of face-to-face and distributed teaching.

11.5.1 Top-down, Bottom-up or "Sideways"?

Top management support, involving a clear managerial mandate for user adoption of the technology, is often emphasised as a critical factor for success of implementation projects. The argument for this is to ensure a steered process, backed by adequate resources and a mandate to implement change. The downside of a top-down approach is of course the danger of forcing the new technology upon the users, unless user participation is emphasised during the process. Thus, many instead argue for a bottom-up process.

Several of the implementation projects in this book actually represent combinations of top-down and bottom-up approaches. As shown in Table 11.2, the initiative and project origin is often traced to lower levels of the organisation, most frequently the IT units. Not until later is the project formalised and implemented using a top-down approach. This was the case for the implementation of Lotus Notes in Statoil. The adoption and diffusion of online calendars in Sun and Microsoft, as described in Chapter 8, is the one closest to a "pure" bottom-up process. After spreading from individual developers to managers and administrative assistants, peer pressure to use the technology was experienced from "every direction" in the company leading to universal adoption. In this process, a managerial mandate was not found to be required for stimulating adoption.

In other cases the importance of explicit management action has been warranted to establish a critical mass of users. This was the case for both GroupSystems in Statoil and KINET in Kværner. Possible strategies for stimulating adoption of a collaboration technology may be to establish some incentives and/or to remove or restrict the use of alternative communications channels or tools. Such actions typically require some form of managerial mandate.

11.5.2 Decentralised Adoption Process

A special form of adoption process can be identified in highly distributed organisations. In these organisations, the decision to adopt a new technology is often decentralised to the organisational unit. Stimulating adoption becomes essentially an internal marketing process, where the implementation team tries to convince individual unit managers to adopt the technology. For example, Statoil IT developed and marketed a Lotus Notes "toolbox" to the different units in the company, who were free to decide which applications from this toolbox they would adopt. And in the implementation of KINET in Kværner, the different companies in the group decided whether to invest in the new technology or not. The process in Kværner was run by the concern management and based

on their vision for improving collaboration and communication. With engineering systems being the main focus in the company, communication technology was not given priority in many companies in the group and this vision thus had to be translated to the local needs as perceived by each company.

11.6 Summary and Conclusion

This chapter has provided an integrated analysis of the findings from implementation projects described in Part 2 of the book. Using the taxonomy of implementation factors introduced in Chapter 4, a cross-case analysis has been presented focusing on the practical experiences from these cases. Findings from the literature review in Chapter 3 have also been drawn upon. This analysis has identified a wide range of factors that have influenced these implementation projects to various degrees. Practical implications from this analysis are presented in the final chapter, in the form of a set of guidelines for implementing collaboration technologies.

12 Implications for Practice and Research

Bjørn Erik Munkvold

12.1 Introduction

The case chapters in Part 2 of the book each present explicit guidelines and recommendations for implementation of different collaboration technologies. These guidelines constitute important takeaways for any reader about to be or already engaged in similar projects. Chapter 11 presented a comparative analysis of the cases, discussing key issues identified from these projects and previous research. This final chapter presents a set of practical guidelines that synthesise the lessons learned from these case studies and from previous field based research covered in Chapter 3. These guidelines address organisational, project-related and technology-related aspects of implementation.

The aim here is not to present a general implementation strategy for collaboration technology. The contextual nature of such implementation processes makes this an unrealistic task. Rather, the purpose is to identify critical factors to address when planning and implementing collaboration technologies. Finally, some implications for research are also discussed.

12.2 Use a Planned Approach – but Prepare for Improvisation

Implementation of collaboration technology requires careful planning. It should be conducted as a project, led by an implementation team. The size and composition of this team needs to be adapted to the scope of the implementation, but it needs to cover business, technical and project management competence. In heterogeneous and/or distributed organisations, the implementation team needs to establish a network of local champions in the different units. These should aid in the marketing and adaptation of the system to the different units, and translating the functionality of the technology to solve local needs.

As illustrated by the case studies in this book, these implementation projects may take several years before the full potential of the technology is utilised. In this process many factors may have an impact on the project, and you often find that conditions change along the way. Also, many such projects have the character of a learning process, where those involved gradually develop an understanding of the features of the technology and how this can be used for supporting different processes and tasks. Thus, it is often not possible to fully predict at the outset what the final outcome of the project will be. Instead, it is important to acknowledge the need for improvisation capability, to overcome barriers encountered and to take advantage of new opportunities opening up.

When initiating this type of project one therefore needs to prepare for an evolutionary process, requiring endurance and patience from those involved. No quick success should be expected. There needs to be some assurance that adequate project resources will be available to back this entire process. This again requires a sponsor for the project. An important means for establishing this sponsorship may be to run one or more pilot projects for demonstrating the potential of the technology. This is further discussed as a separate point below.

12.3 Get to Know the Setting

Successful implementation of collaboration technology requires in-depth knowledge of the organisational and technological context in which the new technology is introduced. This can only be obtained through extensive mapping of the organisational context (culture, norms, incentives, etc.), existing work practices and technological infrastructure. Ideally, this mapping should be based on ethnographic approaches as demonstrated in many field studies in the CSCW area. However, the time and resources required for these approaches may limit their use in projects with tight budgets and strict deadlines. Still, some amount of "fieldwork" is required to get beyond the formal policies and capture the actual work practices. Part of this fieldwork should also focus on identifying important needs of the different user groups and related incentives for adoption and use, to prepare for the local marketing of the technology. In projects where the choice of technological solution is dependent on the users' expressed needs, demonstrations of alternative solutions and their possible application in the actual work context need to be conducted, followed by discussions between designers and users. Otherwise, you cannot expect the users to express needs for technology support for new ways of working of which they are not even aware.

The mapping of organisational context also needs to identify the different stakeholder groups that may be affected by the implementation, and potential conflicts of interest. This type of process often entails a political element and by addressing these issues up front and establishing a discussion related to this, potential conflicts may be solved before they escalate.

12.4 Adapt the Marketing of the Project to the Stakeholder Groups

The implementation of collaboration technology is often based on an overall vision of how the company can benefit from improved communication, coordination and collaboration throughout the organisation. However, as demonstrated in several cases in this book, this corporate "grand vision" does not automatically translate into local incentives for adopting the technology. To stimulate adoption among the different user groups, the marketing needs to focus on how the technology can benefit them in their workday. This again requires knowledge of local work practices and related terminology, and local champions may here make an important contribution.

"What's in it for me?" is the key question that concerns most users. In cases where there are no direct benefits for the individual user, the technology champions need to convince the users how their contribution through adoption and use of the technology will indirectly benefit their organisational unit, through producing benefits for the company as a whole. This is important to overcome the potential feeling of disparity in work and benefit (Grudin, 1994b), where the technology adoption is seen primarily as extra work without immediate or clear benefits. In this process, what may appear as "irrational resistance" from the users should be taken seriously and carefully analysed, to learn about the users' perspectives and provide arguments that may help to change their attitude towards the technology.

The implementation team here needs to act as "missionaries" – especially when the new technology involves a shift to new work practices unknown to the adopters. Demonstrations of prototypes and/or visits to other sites already using this technology can be important. Further, successful pilot projects may aid in convincing the users of the potential of this technology. In general, it is important that all successful experiences with the technology are published through internal information channels (intranets, newsletters, etc.).

Some studies of Lotus Notes implementations show how the technology has been expected to "sell itself", thus not requiring much explicit guidelines and stimulation of adoption beyond installation and initial training of users. However, although rollout and deployment of the technology may function in this way, user adoption of the services seldom does. In general, close follow-up throughout all stages of the implementation is required. This follow-up actually represents a continuous activity for ensuring effective use of the technology.

Finally, in the marketing one should also look for internal or external opportunities and events that may support the adoption of the technology. Examples of this are the focus on reduction of travel costs in Statoil and the merger of Boeing with Rockwell and McDonnell Douglas, both spurring the widespread adoption and use of NetMeeting to support distributed meetings in these organisations.

12.5 Use Pilot Groups

If possible, use one or more pilot groups to learn more about the potential of the technology and demonstrate this for further adoption. The composition of these groups is of key importance – the members of these groups should have a real need for technology support and should also be relatively experienced IT users. In distributed and/or inter-organisational settings, the implementation team needs to participate in selecting pilot team members to ensure that these criteria are met.

In general, the role and function of the pilot group will depend on the intended scope of the full implementation. Opper and Fersko-Weiss (1992) discuss how this scope will affect how the following activities are sequenced when designing a pilot:

- selecting the pilot group – deciding which workgroup will inaugurate collaboration technology first;
- selecting the application – determining the specific business purpose the pilot will be testing;
- selecting the product – finding the right technology.

For enterprise-wide implementations, the product is often selected first, while for smaller scale implementations the product decision can be made last. For staged exploration the sequence can be adapted to the specific context.

12.6 Build a Supportive Infrastructure

The studies in this book have illustrated the importance of establishing a human support infrastructure. Different roles were discussed in the previous chapter: product and project champions, superusers, mediators and facilitators. While some of these roles may emerge without any formal appointment, those in charge of the implementation should be on the lookout for employees that may fill these roles. In any case, these roles should be stimulated and nurtured, through allocating sufficient time away from other tasks and including such positions in the organisation's reward system. Some of these roles also require special training, such as the facilitator role related to the use of electronic meeting systems.

The support infrastructure also includes technical support. This is especially important for collaboration technologies, as the "transaction costs" of switching back to old systems are often low. Having a hotline of dedicated personnel available for continuous technical support is a must. Collaboration technologies may also be used to help provide this support. This may, for example, be in the form of workflow-based helpdesk systems, or using application sharing for troubleshooting and user instructions in distributed settings.

12.7 Invest in Training – Including Collaborative Skills

The importance of user training cannot be overstated. One of the most frequently reported explanations for less successful implementation of collaboration technology is lack of training in the collaborative nature of the technology. Training that is narrowly focused on the "mechanistic" features of the tools, leaves the users to figure the collaborative aspects out for themselves. For collaborative product suites such as Lotus Notes, this has proved especially troublesome, in many cases resulting in Lotus Notes being used as "advanced e-mail" rather than a tool for information sharing. Thus, training should explicitly focus on how this tool can be used *for supporting collaboration.* This type of training in collaborative work practices can actually also be regarded as a way to stimulate the development of a "collaborative culture", that is often emphasised in the literature.

Further, to avoid making this too general, the training needs to be adapted to the individual work tasks of the different user groups. The most effective form of training has an important on-the-job component. Ideally, at least some of the training should take place in relevant situations, at the users' workplace. The development in e-Learning technologies here contributes to greater flexibility in structuring this training, as exemplified by the Terry–PwC Consulting MBA programme presented in Chapter 10.

In the competitive climate facing many organisations, it is often difficult to allocate sufficient time and resources to training in the tight project budgets. As exemplified in the Kværner case, project management here needs to convince the internal or external project owners of the importance of sufficient training for achieving the expected quality and expedience of the project.

In general, implementation of collaboration technologies usually implies new work practices, and explicit training in these should be provided. Of course, some types of collaboration technologies are more intuitive in use than others. For example, the use of asynchronous and synchronous communication tools such as e-mail and videoconferencing does not require as much "conceptual" training as knowledge databases. Still, there is clearly a need for training in how to effectively use these communication services as well.

In general, collaboration requires collaborative skills that cannot be expected to automatically emerge. These skills will vary depending on the form of collaboration. For example, effective teamwork requires skills in teambuilding and team conduct. For virtual teamwork supported by collaboration technology, additional skills are needed. These include ability to develop and maintain interpersonal relations using electronic communications media. In global settings, additional factors such as language and cultural differences, different time zones, etc. come into play. It is important to note how face-to-face teambuilding is emphasised in the early stages of virtual teamwork, such as in the Terry–PwC Consulting programme and in the distributed engineering projects in Kværner.

12.8 Develop Guidelines and Routines for Use

Developing guidelines for effective use is needed for all collaboration technologies. These should be based on a careful analysis of the work setting and how the features of the technology can best be matched with this. However, developing such guidelines requires a period of learning and experimenting with the technology. As experience builds, new ways of utilising the technology may emerge in addition to those originally planned.

Once identified, these guidelines should be formalised as routines and communicated clearly to the users involved as "best practice" for using the technology. All too often, after the initial training courses the users themselves are left with sorting out how to apply the technology to their existing work practices. Especially for organisations using several technologies there is a need to establish routines for how and when to use each service. Still, it is important that these routines do not become straightjackets for the users in their creative problem-solving and task execution. As pointed to in the literature review, there is a need to find the right balance between predefined routines and flexibility for experimentation and improvisation.

It is also important to realise that these guidelines are not developed once and for all, but should be frequently evaluated and adjusted. Thus it is important to stimulate reflection among the users regarding their practices, and also create forums for disseminating and discussing best practice. These forums can both be "physical meetings" and electronic forums where collaboration technology is also used as a vehicle for discussing such practices, for example through using asynchronous tools such as discussion lists and bulletin boards or synchronous discussion through desktop videoconferencing. Also, the facilitator role may have a special responsibility for this activity.

12.9 Managers – Engage and Stimulate!

Top management support is listed as a critical success factor for IT implementation so often that it has nearly become a cliché. Herein lies the danger of underestimating the importance of the *nature* of this support. In several of the implementation projects studied in this book, management involvement and support has been overly passive. Beyond merely approving budget allocations, management has not taken active steps to stimulate adoption or develop new work routines. Often, this is based on lack of knowledge about the technology being implemented.

Managers should take on a proactive role in stimulating adoption and use of the technology among their employees. As concluded in the Boeing case, executive use can be equally powerful as executive sponsorship. This requires that management undertake sufficient training to serve as role models in use of the technology. Second, new incentive systems should be created for stimulating employees' contribution to collaborative processes and forums such as knowledge bases and online discussions. Third, management should participate in developing routines for best practice in use of the technology, and also enforce that these rules are followed. This may involve restricting the use of alternative media for communication and information sharing, and also taking actions if the routines are not followed.

Although not all of these initiatives need to come from management, it is important that they actively support them. Typically, the project champion will take on much of the operationalisation of this, and this person may not be part of the management. However, management is responsible for involving the right group of people in the implementation, including champions, facilitators and other support persons, and creating the best working conditions for these.

12.10 Establish a Formal Organisation for Maintenance and Further Development

Implementation projects of the nature studied in this book are not discrete events with a clear "ending" – rather they develop continuously into new phases. When the technology has become part of the common IT infrastructure in the organisation and is being used regularly, it is important to establish some formal organisation for maintenance and further development of this service. The e-collaboration team in Statoil is an example of an integrated team with responsibility for supporting the entire portfolio of collaboration technologies within the company. Of course, for this solution to be effective requires that competence and focus within the team is equally distributed for the different technologies. In Boeing, this responsibility is handled by the central IS organisation. However, in this case there are dedicated product managers for each technology/service.

12.11 Concluding Remarks

The overall objective of this book is to contribute to an understanding of the many possible factors that may influence the implementation of different collaboration technologies, and present experiences on how different companies have handled this process. This is not a "cookbook" for implementation of collaboration technology – the flexibility in application of collaboration technology combined with contextual variations would make this a rather dubious project. However, the guidelines presented here form the basis for developing implementation strategies adapted to specific projects.

12.12 Implications for Research

Continued research in the field is required to further our understanding of how to effectively implement collaboration technology in different contexts. Although not primarily framed as a research volume, this book contributes to research in several ways. Part 1 of the book presents an overview of collaboration technologies and the current status in field-based research on implementation of collaboration technologies. Through this, several areas are identified where more research is needed. In general, with new technologies emerging, the practices and experiences related to these should be the subject for in-depth studies like the case studies in Part 2 of the book. For example, there are yet few studies of the implementation and use of new web-based collaborative services and mobile applications.

Similarly, as the applications of collaboration technology continue to spread to different sectors and application areas, more case studies are needed on the potential influences and characteristics of these settings. As an example, applications of collaboration technology in the public sector are still at an early stage. There is enormous potential here both for making existing processes more effective and for creating new and exciting forms of public debate and decision-making (for example, related to e-government and teledemocracy).

Several of the findings from the case chapters in Part 2 of the book pave the way for more research. For example, despite the current trend of integration of collaborative tools within products, there is still limited knowledge on how these different tools may best work together. Also, how the implementation of different collaboration technologies within one organisation may be inter-related is an open question. As collaboration technology increasingly is used for linking partners throughout the value chain, new issues emerge related to the ownership of information produced in these collaborative ventures.

Finally, with the increasing focus on global collaboration supported by collaboration technology (as in global virtual teams), one should reflect on the fact that most of the research on collaboration technology to date is based on studies of companies in the Western world. At a global scale, we are only at the brink of utilisation of collaboration technologies. The potential impact of cultural aspects on collaborative work practices creates many interesting questions on how the existing knowledge on collaboration technology adoption and use can be translated to other world regions. These questions should be eagerly pursued by both researchers and practitioners in this field, to contribute to further realisation of the potential benefits of collaboration technology on a global scale.

Appendix A

Collaboration Technology Forums and Resources

This appendix provides an overview of some major forums related to collaboration technology research and practice, as well as some key players among consultants and industry associations/coalitions. The list is not exhaustive, and is only intended to provide the reader with some starting points for further exploration in this field.

The links were operational as of May 2002, and no responsibility is taken for further updates of these.

Major Conferences and Workshops

- CSCW (Computer-Supported Cooperative Work) – bi-annual international conference gathering both industry practitioners and researchers, vendors and academic researchers.
- ECSCW (European Conference on Computer-Supported Cooperative Work) – bi-annual European equivalent to the CSCW conference.
- SIGGROUP – ACM Special Interest Group on Supporting Group Work (www.acm.org/siggroup/)
- HICSS (Hawaii International Conference on System Sciences) – yearly conference held on Hawaii in January. Includes a track for Collaboration Systems and Technology, with several minitracks. The proceedings from this conference is available online from the HICSS Digital Library (www.hicss.org/diglib.htm)

An overview of these and other conferences and forums related to collaboration technology is available from the CSCW 2002 website (www.acm.org/cscw2002/related.html).

Consultants, Forecasting Institutions, Industry Associations and Coalitions

- Bootstrap Alliance (founder: Douglas Englebart) (www.bootstrap.org)

- Collaborative Strategies (www.collaborate.com) – offers free subscription of e-mail newsletter with industry updates on collaboration technologies.
- Gartner Group (www3.gartner.com)
- Institute for the Future (www.iftf.org)
- International Association of Facilitators (www.iaf-world.org)
- The Internet Engineering Task Force (www.ietf.org)
- The Workflow Management Coalition (www.wfmc.org)

Appendix B

Product Examples

This appendix lists some product examples for different categories of collaboration technology. The aim is not to provide a "complete" product listing or present any recommendations. In general, the fast-paced, continuous development in these products makes any "snap-shot" evaluation of these of limited value. Rather, these links are intended to provide the reader with some examples of collaborative product functionality currently available in the market.

The links were operational as of May 2002, and no responsibility is taken for further updates of these.

Instant messaging

- AIM (www.aol.com)
- ICQ (www.icq.com)
- MSN Messenger (messenger.msn.com)
- Odigo (www.odigo.org)
- Yahoo! Messenger (messenger.yahoo.com)

Interactive Whiteboards

- SMART Board (www.smarttech.com)
- LiveBoard (www.wearesimply.com)

Electronic Meeting Systems

- Facilitate.com (http://www.facilitate.com)
- GroupSystems.com (www.groupsystems.com)
- Meetingworks (http://www.meetingworks.com)

Collaboration Product Suites

- Groove (www.groove.net)
- GroupWise (www.novell.com)
- Lotus Notes/Domino (www.lotus.com)
- Microsoft Exchange (www.microsoft.com)
- Teamware (www.teamware.com)

Web-based Project/Teamrooms

- Groove Workspace (www.groove.net)
- Intranets.com (www.intranets.com)
- Lotus Quickplace (www.lotus.com)
- Projectplace.com (www.projectplace.com)

Desktop Conferencing/Real-time Conferencing

- Centra eMeeting (www.centra.com)
- ERoom (www.eroom.com)
- Groove Workspace (www.groove.net)
- Lotus Sametime (www.lotus.com)
- Magi Enterprise (www.endeavors.com)
- MS NetMeeting (www.microsoft.com)
- WebEx Meeting Center (www.webex.com)
- The Thinkofit consultant company offers an extensive list of software and services that enable real-time (synchronous) e-conferencing: http://thinkofit.com/webconf/realtime.htm.

e-Learning Technologies

- Blackboard (www.blackboard.com)
- Centra Symposium (www.centra.com)
- Lotus LearningSpace (www.lotus.com)
- WebEx Training Center (www.webex.com)

Appendix C

A Taxonomy of Implementation Factors for Collaboration Technologies

Implementation factors	Possible effects on implementation
ORGANISATIONAL CONTEXT	
• Existing degree of collaborative work practices in the organisation	Existing collaborative work practices may have a positive effect on the users' receptivity for collaboration technology.
• Users' felt need for technology support	A felt need among the users has a clear positive effect on adoption of the technology.
• Individualistic versus collaborative culture	Organisations with a highly individualistic and competitive culture may face greater challenges in adoption of collaboration technologies than organisations already focusing on collaboration.
• Reward systems and policy	These structural elements are important means for stimulating collaboration and related use of collaboration technology in the company.
• Top management support	Top management support is important for providing organisational "legitimacy" to the implementation and for gaining access to adequate resources.
• Management style	Management style can impact the implementation and use of collaboration technology. However, collaboration technology can be adapted to serve different styles, and does not automatically support more collaborative and decentralised/democratic approaches.
• Existing IT infrastructure	Collaboration technologies require a basic IT infrastructure. The implementation project needs to take into account any necessary upgrades in this infrastructure.

(*cont'd*)

Implementation factors	Possible effects on implementation
• Existing IT competence	Lack of internal IT competence in the organisation may be a barrier to effective implementation. On a short range, vendors and consultants can provide this, but the organisation needs to build internal competence for future maintenance and support.
• Economic conditions	Economic conditions such as recession in national economy and fluctuations in market conjunctures may impact the implementation in different ways. For example, it may result in budget cuts for the implementation, or it may lead to increasing focus on how to make organisational practices more effective through collaboration technology.
IMPLEMENTATION PROJECT	
• Formalised implementation strategy versus improvisation	A formalised implementation strategy has a positive effect on project management, including scheduling, resource allocation and coordination. However, experience shows that some room for improvisation is needed.
• Composition of implementation team	An implementation team with a right blend of technical competence and business understanding creates the required "socio-technical balance" needed for successful implementation.
• Information to the users	The information provided to the users has an important bearing on their perceptions (mental models) of the collaboration technology and its potential.
• Users' expectations	Realistic expectations towards the new technology are important to avoid any frustration and disappointment among the users. Potential benefits of the technology should be communicated to the users, but without "overselling" it.
• Composition of pilot groups	Pilot groups without a real need for technology support may fail to document the potential benefits. The members of the pilot groups should be selected on the basis of their need for collaborative IT support.
• User training	Lack of adequate training is a recurring factor in implementation failure. The training needs to include an explicit focus on collaborative aspects.
• Establishing a supportive infrastructure	Some form of support infrastructure is important to handle problems early and thus avoid user frustration.

(cont'd)

Implementation factors	Possible effects on implementation
• Project champion(s)	Access to one or more project champions has proven instrumental to implementation success.
• Incentives for stimulating user adoption	Establishing clear incentives may stimulate adoption of the technology. This could be in the form of improved working conditions for the individual employees, and/or bonus schemes for increased productivity.
• Predefined routines versus user experimentation	Clear guidelines and routines may increase the effect of the technology. This should be balanced against giving the users room to experiment with the technology, to come up with new and creative applications.
GENERAL TECHNOLOGY-RELATED FACTORS	
• Critical mass	Establishing a critical mass of users is crucial for collaboration technologies where the users' benefits are dependent on universal adoption.
• Disparity in work and benefit	Perceived disparity in extra workload and benefit induced from the technology may represent a barrier to user adoption.
• Disruption of social processes	Technologies that represent disturbances to the often tacit social processes risk facing user resistance.
• Exception handling	Exceptions to the formal routines occur frequently in the day-to-day work practices. Some flexibility should be built into the systems, to accommodate for these exceptions.
• Unobtrusive accessibility	Some collaborative tools are not used as frequently as other office support tools. By offering seamless integration with the user's standard work tools, the collaboration tools also accommodate more infrequent use.
• IT maturity	Immature technology can create problems with stability and performance of the solution, resulting in project delays and distrust among the users.
• Compatibility with existing technologies	Technical incompatibility can result in project delays and frustrated users.
• Compatibility with existing routines	Compatibility with existing routines means less "friction" in user adoption. However, some implementations will aim at changing these routines.

(cont'd)

Implementation factors	Possible effects on implementation
• Fragile nature of collaboration technologies	In case of problems with a new collaboration technology, users may easily abandon this in favour of existing, substitute technologies more familiar to them.
SPECIFIC TECHNOLOGY-RELATED FACTORS	
COMMUNICATION TECHNOLOGIES	
• Routines for electronic communication	Such routines may contribute to effective use of the services, and reduced information overload.
• Social protocols for communication frequency and contents (netiquette)	Important for building relationships in electronic communication, and avoiding misbehaviour.
• Bandwidth and image quality	Critical factors for videoconferencing systems.
SHARED INFORMATION SPACE TECHNOLOGIES	
Document management systems/Knowledge repositories	
• Increasing visibility of document production process	Transition from paper-based to electronic document handling makes the document production process more transparent. This requires an analysis of possible changes in the temporality of work routines, such as related to publication and distribution of documents.
• Ownership and responsibility of information	Sharing electronic documents may raise new issues related to ownership and responsibility for the information in its various production stages.
• Support for mobile users	Limited access to digital documents for mobile users may represent a challenge in the implementation of these technologies.
• Effective search mechanisms	Critical for effective use of knowledge repositories. For organisation-wide databases, there may be a need for developing a thesaurus.
• Roles and responsibilities for content management	Effective content management using document management systems requires new roles and responsibilities for maintenance and quality control.
Data conferencing/Application sharing	
• Distributed facilitators	There is a growing attention to the importance of this role to ensure effective communication "flow" in distributed meetings.
• Technical support	Dedicated technical support can eliminate start-up delays and problems in distributed meetings. This function can also be fulfilled by the distributed facilitators. *(cont'd)*

Implementation factors	Possible effects on implementation
• Routines/protocols for structured use of application sharing (screen management and turn-taking)	Such routines/protocols are needed to avoid "chaos", and ineffective use of application sharing.
• Audio quality	Limited audio quality may restrict the use of integrated audio and data conferencing.
ELECTRONIC MEETING SYSTEMS	
• Dedicated electronic meeting rooms	Co-located, electronic meetings require dedicated meeting rooms with adequate IT infrastructure. This may be a significant investment for a company.
• Access to trained facilitators	The meeting facilitator is instrumental for successful electronic meetings. He or she is responsible for planning and running the meeting, and processing the meeting report.
• Matching EMS tools with meeting tasks	Using the right EMS tools for the meeting activities is vital for the process and outcome of an electronic meeting. This is specified in the meeting agenda prepared by the facilitator.
• Balancing electronic and verbal interaction	Electronic meetings require a balance of electronic and verbal interaction to be effective. The facilitator manages this balance.
COORDINATION TECHNOLOGIES	
Workflow management systems	
• Transferring business process model into workflow model	The workflow model needs to incorporate both organisational and technical aspects of the business process.
• Correspondence with users' model of work	Imposing new work models that do not correspond with the users' model of work may disrupt the "smooth flow of work" and lead to user resistance.
• Flexibility in process	Necessary for exception handling and allowing some user autonomy in job allocation and prioritisation.
• Timing of selection of workflow product	The workflow product should not be selected until after the new business process model has been designed, to assure that the product meets the requirements in full.
• Integration with legacy systems	Important but often challenging and resource demanding task in workflow implementation.

(*cont'd*)

Implementation factors	Possible effects on implementation
• Management surveillance	Potential risk of misuse for control purposes may result in users being sceptic. It is important to deal with this up front, to reassure users.
INTEGRATED PRODUCTS	
Lotus Notes	
• Users' individual interpretations (mental models) of the technology	The users' interpretations of the technology frame the scope and effectiveness of its use. Explicit training in the collaborative features of the technology is important for demonstrating its potential to the users.
• Balance between management directives and user experimentation and improvisation	Some guidelines for "best practice" are needed to ensure effective use. This must be balanced against the need for allowing users to experiment with the technology to come up with creative applications.
IMPLEMENTATION PROCESS	
• Top-down versus bottom-up approach	A top-down implementation approach may ensure a coordinated process guided by an overall vision, but may face user resistance due to lack of adaptation to local needs and practices. A bottom-up approach may result in greater "buy in" from the users, but may lack coordination and strategic vision. When possible, a "combined approach" is recommended, stimulating bottom-up adoption guided by strategic vision and central coordination.
• Social influence mechanisms	Social influence mechanisms such as peer pressure and "word of mouth" can be more influential on user adoption of a new technology than any planned approach. The implementation team should try to capitalise on this through appointing superusers and "technology ambassadors" in the organisation.
• Implementation barriers resulting from conflict between organisational context and technology characteristics	Most implementation projects encounter unforeseen barriers threatening the project. The implementation team must deal with these as early as possible, and try to eliminate any misfit between technology and organisational context.
• User learning and adaptation	The implementation cases show that users generally are able to adapt to changing work practices and use of new collaboration technology – however, this is a gradual learning process that may take long.

References

Abdel-Wahab, H.M. and Feit, M.A. (1991). XTV: A Framework for Sharing X Window Clients in Remote Synchronous Collaboration. In *Proceedings of IEEE Conference on Communications Software*, IEEE Communications Society, 159–167.

Akselsen, S. (1999). Distributed Teams and the Individual: ICTs for Dealing with Challenges of New Work Arrangements. *Telektronikk*, 4/99. Kjeller, Telenor R&D, Norway.

Akselsen, S., Bergersen, E., Svendsen, G. and Folkow, T. (1996). *Markedsinnovasjon – Første kokebok*. Report 27/96. Kjeller, Telenor R&D, Norway. (In Norwegian).

Anson, R., Bostrom, R. and Wynne, B. (1995). An Experiment Assessing Group Support System and Facilitator Effects on Meeting Outcomes. *Management Science*, Vol. 41, No. 2, 190–208.

Anson, R. and Munkvold, B.E. (2002). Beyond Face-to-face: A Field Study of Electronic Meetings in Different Time and Place Modes. Forthcoming in *Journal of Organizational Computing and Electronic Commerce*.

Argyris, C. and Schön, D.A. (1990). Participatory Action Research and Action Science Compared: A Commentary. In W.F. Whyte (ed.), *Participatory Action Research*, Sage, Newbury Park, CA, 85–96.

Atkinson, D.J. and Lam, A. (1999). A Case Study Exploration of Groupware Supported Workflow. *Australian Computer Journal*, Vol. 31, No. 4, 124–130.

Beard, D., Palaniappan, M., Humm, A., Banks, D. and Nair, A. (1990). A Visual Calendar for Scheduling Group Meetings. *Proceedings of CSCW '90*, ACM, New York, 279–290.

Benbasat, I. and Zmud, R.W. (1999). Empirical Research in Information Systems: The Practice of Relevance. *MIS Quarterly*, Vol. 23, No. 1, 3–16.

Bikson, T.K. and Eveland, J.D. (1996). Groupware Implementation: Reinvention in the Sociotechnical Frame. *Proceedings of CSCW '96*, Cambridge, MA, 428–437.

Blundell, D. (1997). Collaborative Presentation Technologies: Meetings, Presentations, and Collaboration. In Coleman, D. (ed.), *Groupware: Collaborative Strategies for Corporate LANs and Intranets*. Prentice Hall, Upper Saddle River, NJ, 269–318.

Bond, B., Burdick, D., Miklovic, D., Pond, K. and Eschinger, C. (1999). C-Commerce: The New Arena for Business Applications. *Research Note*, Gartner Group.

Bostrom, R.P. (1983). Designing an Information System: The Socio-Technical Approach. *Cause/Effect*, Vol. 6, No. 2, 22–26.

Bowers, J. (1994). The Work to Make a Network Work: Studying CSCW in Action. *Proceedings of CSCW '94*, ACM Press, Chapel Hill, NC, 287–299.

Bowers, J., Button, G. and Sharrock, W. (1995). Workflow from Within and Without: Technology and Cooperative Work on the Print Industry Shopfloor. *Proceedings of ECSCW '95*, Stockholm, 51–66.

Bratteteig, T. (1998). The Unbearable Lightness of Grouping: Problems of Introducing Computer Support for Cooperative Work. *Proceedings of NOKOBIT '98*, Oslo, 99–113.

Bullen, C.V. and Bennett, J.L. (1990). Learning from User Experience with Groupware. *Proceedings of CSCW '90*, Los Angeles, ACM Press, 291–302.

Butterfield, J., Rathnam, S. and Whinston, A. B. (1993). Groupware Perceptions and Reality: An E-mail Survey. *Proceedings of the 26th Annual Hawaii International Conference on System Sciences (HICSS '26)*, 208–217.

Candler, J.W., Palvia, P.C., Thompson, J.D. and Zeltmann, S.M. (1996). The Orion Project: Staged Business Process Reengineering at FedEx. *Communications of the ACM*, Vol. 39, No. 2, 99–197.

Ciborra, C.U. (ed.) (1996). *Groupware and Teamwork: Invisible Aid or Technical Hindrance?* Wiley, Chichester.

Cockshoot, D. (2000). High-value Engineering Using B2B. *Kvaerner E&C Bulletin*, Issue 5, August 2000. Available online at http://www.kvaerner.com/eandc/bulletin/archive/issue-05/feat/feat03.html.

Collaborative Strategies (2000). *The Real Time Collaboration Industry Report 2000*, San Francisco.

Cooper, R.B. and Zmud, R.W. (1990). Information Technology Implementation Research: A Technological Diffusion Approach. *Management Science*, Vol. 36, No. 2, 123–139.

Cummings, T.G. and Worley, C.G. (2001). *Essentials of Organization Development and Change.* South-Western College Publications, Cincinnati, OH.

DataBeam (1997). A Primer on the T.120 Series Standard. *DataBeam Corporation White Paper.* Available online from http://www.lotus.com/products/learnspace.nsf/95febc158ad081518525674c0065aa5f/cb3a67f9f86a93898525679900531dc7?OpenDocumen.

DataBeam (1998). A Primer on the H.323 Series Standard (version 2). *DataBeam Corporation White Paper.*

Davenport, T.H. and Prusak, L. (1998). *Working Knowledge. How Organizations Manage What They Know.* Harvard Business School Press, Boston, MA.

Day-Ryan, P. (1997). *Motorola Net.120 Usability Study.* Motorola Internal Report.

Dennis, A.R., Wixom, B.H. and Vandenberg, R.J. (2001). Understanding Fit and Appropriation Effects in Group Support Systems via Meta-Analysis. *MIS Quarterly*, Vol. 25, No. 2, 167–193.

DeSanctis, G. and Gallupe, R.B. (1987). A Foundation for the Study of Group Decision Support Systems. *Management Science*, Vol. 33, No. 5, 3–10.

DeSanctis, G and Poole, M.S. (1994). Capturing the Complexity in Advanced Technology Use: Adaptive Structuration Theory. *Organization Science*, Vol. 5, No. 2, 121–147.

DeSanctis, G., Poole, M.S, Dickson, G.W. and Jackson, B.M. (1993). An Interpretive Analysis of Team Use of Group Technologies. *Journal of Organizational Computing*, Vol. 3, No. 1, 1–29.

Downing, C.E. and Clark, A.S. (1999). Groupware in Practice: Expected and Realized Benefits. *Information Systems Management*, Vol. 16, Issue 2, 25–31.

Egido, C. (1988). Videoconferencing as a Technology to Support Group Work: A Review of its Failure. *Proceedings of CSCW '88*, ACM Press, Portland, OR, 13–23.

Ehrlich, S.F. (1987a). Social and Psychological Factors Influencing the Design of Office Communication Systems. *Proceedings of CHI+GI '87*, ACM, New York, 323–329.

Ehrlich, S.F. (1987b). Strategies for Encouraging Successful Adoption of Office Communication Systems. *ACM Transactions on Office Information Systems*, Vol. 5, No. 4, 340–357.

Elden, M. (1979). Three Generations of Worker Democracy Research in Norway. In C.L. Cooper and E. Mumford (eds.), *The Quality of Work Life in Europe*, Associate Business Press, London.

Elden, M and Chisholm, R. (1993). Emerging Varieties of Action Research: Introduction to the Special Issue. *Human Relations*, Vol. 46, No. 2, 121–142.

Elden, M and Levin, M. (1990). Cogenerative Learning: Bringing Participation into Action Research. In W.F. Whyte (ed.), *Participatory Action Research*, Sage, Newbury Park, CA, 127–142.

Ellis, C.A., Gibbs, S.J. and Rein, G.L. (1991). Groupware: Some Issues and Experiences. *Communications of the ACM*, Vol. 34, No. 1, 39–58.

Engelbeck, G. and Poltrock, S.E. (1998). Designing to Support Enterprise-wide Collaboration, *Proceedings of the Third International Conference on the Design of Cooperative Systems*, Vol. II, 37–39.

Evaristo, R. and Munkvold, B.E. (2002). Collaborative Infrastructure Formation in Virtual Projects. *Journal of Global Information Technology Management*, Vol 5, No. 2, 29–47.

Evjemo, B., Grav, J., Akselsen, S., Bergvik, S. and Stenvold, L.A. (1999). *Out of Sight – Out of Mind: A Longitudinal Study into an ICT Supported Distributed Group.* Report 34/99. Kjeller, Telenor R&D, Norway.

Finholt, T.A., Rocco, E., Bree, D., Jain, N. and Herbsleb, J.D. (1999). NotMeeting: A Field Trial of NetMeeting in a Geographically Distributed Organization. *SIGGROUP Bulletin*, Vol. 20, No. 1, 66–69.

Finholt, T. and Sproull, L. (1990). Electronic Groups at Work. *Organization Science*, Vol.1, No. 1, 1990, 41–64.

Finn, K.E., Sellen, A.J. and Wilbur, S.B. (eds.). (1997). *Video-Mediated Communication.* Lawrence Erlbaum Associates, New Jersey.

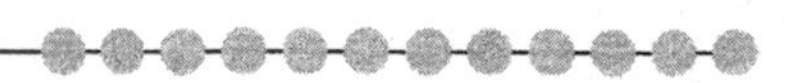

Fjermestad, J. and Hiltz, S.R. (1999). An Assessment of Group Support Systems Experimental Research: Methodology and Results. *Journal of Management Information Systems*, Vol. 15, No. 3, 7–149.

Fjermestad, J. and Hiltz, S.R. (2001). Group Support Systems: A Descriptive Evaluation of Case and Field Studies. *Journal of Management Information Systems*, Vol. 17, No. 3, 115–159.

Francik, E., Rudman, S.E., Cooper, D. and Levine, S. (1991). Putting Innovation to Work: Adoption Strategies for Multimedia Communication Systems. *Communications of the ACM*, Vol. 34, No. 12, 52–63.

Garfinkel, D., Gust, P., Lemon, M. and Lowder, S. (1989). *The Shared X Multi-User Interface User's Guide*, Version 2.0. Hewlett Packard Laboratories.

Ginzberg, M.J. (1978). Steps towards More Effective Implementation of MS and MIS. *Interfaces*, Vol. 8, No. 3, 57–63.

Greenberg, S. and Roseman, M. (1999). Groupware Toolkits for Synchronous Work. In M. Beaudouin-Lafon (Ed.), *Computer supported co-operative work*, John Wiley & Sons, Chichester, 135–168.

Grinter, B. (2000). Workflow Systems: Occasions for Success and Failure. *Computer Supported Cooperative Work (CSCW)*, Vol. 9, No. 2, 189–214.

Grohowski, R., McGoff, C., Vogel, D., Martz, B. and Nunamaker, J. (1990). Implementing Electronic Meeting Systems at IBM: Lessons Learned and Success Factors. *MIS Quarterly*, Vol. 14, No. 4, 369–82.

Grudin, J. (1988). Why CSCW Applications Fail: Problems in the Design and Evaluation of Organizational Interfaces. *Proceedings of CSCW '88*, 85–93.

Grudin, J. (1989). Why Groupware Applications Fail: Problems in Design and Evaluation. *Office: Technology and People*, Vol. 4, No. 3, 245–264.

Grudin, J. (1993). Two Communities, Two Languages. *Communications of the ACM*, Vol. 36, No. 4, p. 113.

Grudin, J. (1994a). Computer-Supported Cooperative Work: History and Focus. *IEEE Computer*, Vol. 27, No. 5, 19–25.

Grudin, J. (1994b). Groupware and Social Dynamics: Eight Challenges for Developers. *Communications of the ACM*, Vol. 37, No. 1, 92–105.

Grudin, J. and Palen, L. (1995). Why Groupware Succeeds: Discretion or Mandate? *Proceedings of ECSCW '95*, Kluwer, Dordrecht, 263–278.

Grudin, J. and Poltrock, S. E. (1990). Computer Supported Cooperative Work and Groupware. *CHI'90 Tutorial Notes*, ACM, New York.

Grudin, J. and Poltrock, S.E. (1997). Computer-Supported Cooperative Work and Groupware. In M. Zelkowitz (ed.), *Advances in Computers*, Vol. 45, 269–320.

Herbsleb, J.D., Mockus, A., Finholt, T. A. and Grinter, R.E. (2000). Distance, Dependencies, and Delay in a Global Collaboration. *Proceedings of CSCW 2000*, Philadelphia, 319–328.

IETF (1995). Netiquette Guidelines. RFC 1855, *The Internet Engineering Task Force*. Available online from http://www.ietf.org/rfc/rfc1855.txt.

IETF (2002). Calendaring and Scheduling (calsh). Working Group on Calendaring and Scheduling, *The Internet Engineering Task Force*. Available online from http://www.ietf.cnri.reston.va.us/html.charters/calsch-charter.html.

Jessup, L.M. and Valacich, J.S. (eds.) (1993). *Group Support Systems: New Perspectives*. Macmillan, New York.

Johansen, R. (ed.). (1988). *Groupware: Computer Support for Business Teams*. Free Press, New York.

Johnson, D. (2001). Special Report: Learning – Next Frontiers. *Newsweek*, 29 October.

Johnson, S. and Blanchard, K. (1998). *Who Moved my Cheese? An Amazing Way to Deal with Change in your Work and in your Life*. G.P. Putnam's, New York.

Johnson, D.W., Johnson, R.T. and Smith, K.A. (1998). Cooperative Learning Returns to College. *Change*, July/August 1998, 27–35.

Karsten, H. (1999). Collaboration and Collaborative Information Technologies: A Review of the Evidence. *The DATA BASE for Advances in Information Systems*, Vol. 30, No. 2, 44–65.

Karsten, H., and Jones, M. (1998). The Long and Winding Road: Collaborative IT and Organisational Change. *Proceedings of CSCW '98*, Seattle, 29–38.

Kautz, K. (1996). Information Technology Transfer and Implementation: The Introduction of an Electronic Mail System in a Public Service Organization. In Kautz, K. and Pries-Heje, J. (eds.), *Diffusion and Adoption of Information Technology*, Chapman & Hall, London, 83–95.

Knudsen, C. and Wellington, D. (1997). Calendaring and Scheduling: Managing the Enterprise's Most Valuable, Non-Renewable Resource – Time. In Coleman, D. (ed.), *Groupware: Collaborative Strategies for Corporate LANs and Intranets*. Prentice Hall, Upper Saddle River, NJ, 115–142.

Kock, N. (1999). *Process Improvement and Organizational Learning: The Role of Collaboration Technologies*. Idea Group Publishing, Hershey, PA.

Korpela, E. (1994). What is Lotus Notes? – An Interpretive Study of Individual and Shared Images of Groupware. In P. Kerola et al. (eds.), *Proceedings of the 17th IRIS*, University of Oulu, Finland, 463–481.

Kotter, J.P. (1995). Leading Change: Why Transformational Efforts Fail. *Harvard Business Review*, Vol. 73, No. 2, 59–65

Kraut, R., Sumais, S., Koch, S. and Kling, R. (1989). Computerization, Productivity and Quality of Work Life. *Communications of the ACM*, Vol. 32, No. 2, 220–238.

Kwon, T.H. and Zmud, R.W. (1987). Unifying the Fragmented Models of Information Systems Implementation. In R.J. Boland and R.A. Hirschheim (eds.), *Critical Issues in Information Systems Research*. Wiley and Sons, Chichester, 227–251.

Line, L. (1998). Application Sharing: A Key Service for a Distributed Organisation. *Proceedings of CIB W78 Conference*, Stockholm.

Lloyd, P. and Whitehead, R. (1996). *Transforming Organizations through Groupware. Lotus Notes in Action*. Springer, London.

Luciano, C., Banerjee, P. and Mehrotra, S. (2001). 3D Animation of Telecollaborative Anthropomorphic Avatars. *Communications of the ACM*, Vo. 44, No. 12, 64–67.

Mackay, W. (1991). Triggers and Barriers to Customizing Software. *Proceedings of CHI'91*, ACM, New Orleans, 153–160.

Malone, T.W. and Crowston, K. (1990). What is Coordination Theory and How Can it Help Design Cooperative Work Systems? *Proceedings of CSCW '90*, Los Angeles, 357–369.

Mark, G., Grudin, J. and Poltrock, S. E. (1999). Meeting at the Desktop: An Empirical Study of Virtually Collocated Teams. *Proceedings of ECSCW'99*, Copenhagen, Denmark, 159–178.

Mark, G. and Poltrock, S. (2001). Diffusion of a Collaborative Technology Across Distance. *Proceedings of the ACM 2001 International Conference on Supporting Group Work (Group '01)*, Boulder, CO, 232–241.

Markus, M.L. (1987). Toward a "Critical Mass" Theory of Interactive Media. Universal Access, Interdependence and Diffusion. *Communication Research*, Vol. 14, No. 5, 491–511.

Markus, M.L. (1995). Disimpacting Use: How Use of Information Technology Creates and Sustains Organizational Transformation. *Working paper*, February 1995.

Markus, M.L. (1997). The Qualitative Difference in Information Systems Research and Practice. In A.S. Lee et al. (eds.), *Information Systems and Qualitative Research*, Chapman & Hall, London, 11–27.

Markus, M.L. and Connolly, T. (1990). Why CSCW Applications Fail: Problems in the Adoption of Interdependent Work Tools. *Proceedings of CSCW '90*, Los Angeles, ACM Press, 371–380.

Marshak, R.T. (1997). Workflow: Applying Automation to Group Processes. In Coleman, D. (ed.), *Groupware: Collaborative Strategies for Corporate LANs and Intranets*. Prentice Hall, Upper Saddle River, NJ, 143–182.

McGrath, J.E. and Hollingshead, A.B. (1994). *Groups Interacting With Technology*, Vol. 194. Sage Library of Social Research, SAGE Publications Inc.

Monteiro, E. and Hepsø, V. (2000). Infrastructure Strategy Formation: Seize the Day at Statoil. In C. Ciborra (ed.), *From Control to Drift: The Dynamics of Corporate Information Infrastructure*. Oxford University Press, 148–171.

Moore, C. and Jones, M. (2001). Comdex: E-learning Touted as Next Killer App, *Computerworld*, November 15, 2001. Available online from http://www.computerworld.com.

Munkvold, B.E. (1998a). Implementation of Information Technology for Supporting Collaboration in Distributed Organizations. *Dr.ing. thesis 1998:40*, The Norwegian University of Science and Technology, Department of Industrial Economics and Technology Management, Trondheim, Norway.

Munkvold, B.E. (1998b). Adoption and Diffusion of Collaborative Technology in Inter-organizational Networks. *Proceedings of the Thirty-First Annual Hawaii International Conference on System Sciences, Vol I*, IEEE Computer Society, 424–433.

Munkvold, B.E. (1999). Challenges of IT Implementation for Supporting Collaboration in Distributed Organizations. *European Journal of Information Systems 8*, 260–272.

Munkvold, B.E. (2000). Alignment of Collaboration Technology Adoption and Organizational Change: Findings from Five Case Studies. In Khosrowpour, M. (ed.), *Challenges of Information Technology Management in the 21st Century*. Idea Group Publishing, Hershey, USA, 601–604.

Munkvold, B.E. and Anson, R. (2001). Organizational Adoption and Diffusion of Electronic Meeting Systems: A Case Study. *Proceedings of the ACM 2001 International Conference on Supporting Group Work (Group'01)*, Boulder, CO, 279–287.

Nardi, B., Whittaker, S. and Bradner, E. (2000). Interaction and Outeraction: Instant Messaging in Action. *Proceedings of CSCW 2000*. Philadelphia, PA, ACM, 79–88.

Nunamaker, J.F. and Briggs, R.O. (1997). Lessons from a Dozen Years of Group Support Systems Research: A Discussion of Lab and Field Findings. *Journal of Management Information Systems*, Vol. 13, No. 3, 163–205.

Nymo, B. (1993). Special Issue on Telemedicine. *Telektronikk*, Vol. 89, No. 1, Kjeller, Telenor R&D, Norway, 42–47.

Okamura, K., Fujimoto, M., Orlikowski, W.J. and Yates, J. (1994). Helping CSCW Applications Succeed: The Role of Mediators in the Context of Use. *Proceedings of CSCW '94*, ACM, New York, 55–65.

Opper, S. and Fersko-Weiss, H.F. (1992). *Technology for Teams: Enhancing Productivity in Networked Organizations*. Van Nostrand Reinhold, New York.

Orlikowski, W.J. (1992). Learning from Notes: Organizational Issues in Groupware Implementation. *Proceedings of CSCW '92*, Toronto, 362–369.

Orlikowski, W.J. (1996a). Evolving with Notes: Organizational Change around Groupware Technology. In C.U. Ciborra (ed.), *Groupware and Teamwork: Invisible Aid or Technical Hindrance?* John Wiley & Sons, Chichester, pp. 23–59.

Orlikowski, W.J. (1996b). Improvising Organizational Transformation over Time: A Situated Change Perspective. *Information Systems Research*, Vol. 7, No. 1, 63–92.

Orlikowski, W.J. and Hofman, J.D. (1997). An Improvisational Model for Change Management: The Case of Groupware Technologies. *Sloan Management Review*, Winter 1997, 11–21.

Palen, L. (1998). *Calendars on the New Frontier: Challenges of Groupware Technology*. Dissertation, Information & Computer Science, University of California, Irvine.

Palen, L. (1999). Social, Individual and Technological Issues for Groupware Calendar Systems. *Proceedings of CHI '99*, 17–24.

Perey, C. (1997). Desktop Videoconferencing. In Coleman, D. (ed.), *Groupware: Collaborative Strategies for Corporate LANs and Intranets*. Prentice Hall, Upper Saddle River, NJ, 321–343.

Pettigrew, A.M. (1990). Longitudinal Field Research on Change: Theory and Practice. *Organization Science*, Vol. 1, No. 3, 267–292.

Poltrock, S.E. (1996). *Some Groupware Challenges Experienced at Boeing*. Paper presented at a workshop of CSCW 96. Available online from http://orgwis.gmd.de/~prinz/cscw96ws/poltrock.html.

Poltrock, S.E. and Engelbeck, G. (1999). Requirements for a Virtual Collocation Environment, *Information and Software Technology*, Vol. 41, 331–339.

Post, B.Q. (1992). Building the Business Case for Group Support Technology. *Proceedings of HICSS 1992*, Hawaii, 34–45.

Prinz, W., Mark, G. and Pankoke-Babatz, U. (1998). Designing Groupware for Congruency in Use. *Proceedings of CSCW'98*, Seattle, Washington, 373–382.

Ragusa, J.M. and Bochenek, G.M. (2001). Collaborative Virtual Design Environments. *Communications of the ACM*, Vol. 44, No. 12, 41–43.

Rocco, E., Finholt, T.A., Hofer, E.C., and Herbsleb, J.D. (2001). Out of Sight, Short of Trust. *Proceedings of the European Academy of Management*, 19–21 April 2001, Barcelona.

Rogers, E.M. (1983). *Diffusion of Innovations. Third Edition*. The Free Press, New York.

Rogers, E.M. (1995). *Diffusion of Innovations. Fourth Edition*. The Free Press, New York.

Ruttenbur, B.W., Spickler, G.C. and Lurie, S. (2000). eLearning: The Engine of the Knowledge Economy. *Morgan Keegan & Co. Inc.*

Sanderson, D. (1992). The CSCW Implementation Process: An Interpretative Model and Case Study of the Implementation of a Videoconference System. *Proceedings of CSCW '92*, Toronto, 370–377.

Schrum, L. and Benson, A. (2000a). A Case Study of One Online MBA Program: Lessons from the First Iteration of an Innovative Educational Experience. *The Business, Education and Technology Journal*, Vol. 2, No. 2, 38–46.

Schrum, L. and Benson, A. (2000b). Online Professional Education: A Case Study of an MBA Program through its Transition to an Online Model. *Journal of Asynchronous Learning Environments* [On-line serial], Vol. 4, No. 1. Available online from http://www.aln.org/alnweb/journal/jaln-vol4issue1.htm.

Schön, D.A. (1963). Champions for Radical New Inventions. *Harvard Business Review*, March-April, 77–86.

Smith, R.C. (2001). Shared Vision. *Communications of the ACM*, Vo. 44, No. 12, 45–48.

"Special Report: E-Learning" (2001). *US News and World Report*, 15 October 2001.

Sproull, L.S. and Kiesler, S. (1986). Reducing Social Context Cues: Electronic Mail in Organizational Communication. *Management Science*, Vol. 32, No. 11, 1492–1512.

Sproull, L.S. and Kiesler, S. (1991). *Connections: New Ways of Working in the Networked Organization*. MIT Press, Cambridge, MA.

Stenvold, L.A, Svendsen, G., Folkow, T. and Akselsen, S. (1997). A Method for Creating and Introducing New Telecom Services and Products. *Human Factors in Telecommunication – HFT'97*, Oslo, 12–16 May. Available online from http://www.tft.tele.no/sigmund/work/hft97/hft97.htm.

Stohr, E.A. and Zhao, J.L. (2001). Workflow Automation: Overview and Research Issues. *Information Systems Frontiers*, Vol. 3, No. 3, 281–296.

Sund, T., Rinde, E. and Størmer, J. (1991). *The TMS teleradiology Experiment*. Report 12/91. Kjeller, Telenor R&D, Norway.

Thorsrud, E. (1977). Democracy at Work: Norwegian Experiences with Non-Bureaucratic Forms of Organization. *Applied Behavioral Science*, Vol. 13, No. 3, 410–421.

Tyran, C.K. and Dennis, A.R. (1992). The Application of Electronic Meeting Technology to Support Strategic Management. *MIS Quarterly*, Vol. 16, No. 3, 313–353.

Vandenbosch, B. and Ginzberg, M.J. (1997). Lotus Notes® and Collaboration: Plus ça change... *Journal of Management Information Systems*, Winter 1996–97, Vol. 13, No. 3, 65–81.

Votsch, V. (2001). A Taxonomy for Content Management Systems. *The Seybold Report*, Vol. 1, No. 11, 13–19.

Walsham, G. (1993). *Interpreting Information Systems in Organizations*. John Wiley and Sons, Chichester.

Watson, R.T., Akselsen, S., Evjemo, B. and Aarsæther, N. (1999). Teledemocracy in Local Government. *Communications of the ACM*, Vol. 42, No. 12, 58–63.

Weske, M., Goesmann, T., Holten, R. and Striemer, R. (1999). A Reference Model for Workflow Application Development Processes. *Proceedings of WACC '99*, San Francisco, 1–10.

WfMC (2002). Workflow Management Coalition. http://www.wfmc.org/.

White, D. (2000). Why B2B is Changing the Way We Do Business. *Kvaerner E&C Bulletin*, Issue 5, August. Available online from http://www.kvaerner.com/eandc/bulletin/archive/issue-05/feat/feat02.html.

Whyte, W.F. (ed.) (1990). *Participatory Action Research*. Sage, Newbury Park, CA.

Woitass, M. (1990). Coordination of Intelligent Office Agents: Applied to Meeting Scheduling. In S. Gibbs and A.A. Verrijn-Stuart (eds.), *Multi-User Interfaces and Applications*, North Holland, Amsterdam, 371–387.

Yin, R K. (1994). *Case Study Research*. Sage Publications, Newbury Park, CA.

Yourdon, E. (1996). When Good Enough Is Best. *BYTE*, September, 85–90.

Ytterstad, P., Akselsen, S., Svendsen, G. and Watson, R. (1996). Teledemocracy: Using Information Technology to Enhance Political Work. *MIS Quarterly*, Vol. 20, No. 3, Available online from MIS Quarterly Discovery at http://www.misq.org/discovery/articles96/article1/

Index